Springer Finance

Editorial Board
M. Avellaneda
G. Barone-Adesi
M. Broadie
M.H.A. Davis
E. Derman
C. Klüppelberg
E. Kopp
W. Schachermayer

Springer Finance

Springer Finance is a programme of books aimed at students, academics and practitioners working on increasingly technical approaches to the analysis of financial markets. It aims to cover a variety of topics, not only mathematical finance but foreign exchanges, term structure, risk management, portfolio theory, equity derivatives, and financial economics.

M. Ammann, Credit Risk Valuation: Methods, Models, and Application (2001)
K. Back, A Course in Derivative Securities: Introduction to Theory and Computation (2005)
E. Barucci, Financial Markets Theory. Equilibrium, Efficiency and Information (2003)
T.R. Bielecki and M. Rutkowski, Credit Risk: Modeling, Valuation and Hedging (2002)
N.H. Bingham and R. Kiesel, Risk-Neutral Valuation: Pricing and Hedging of Financial Derivatives (1998, 2nd ed. 2004)
D. Brigo and F. Mercurio, Interest Rate Models: Theory and Practice (2001)
R. Buff, Uncertain Volatility Models-Theory and Application (2002)
R.A. Dana and M. Jeanblanc, Financial Markets in Continuous Time (2002)
G. Deboeck and T. Kohonen (Editors), Visual Explorations in Finance with Self-Organizing Maps (1998)
R.J. Elliott and P.E. Kopp, Mathematics of Financial Markets (1999, 2nd ed. 2005)
H. Geman, D. Madan, S. R. Pliska and T. Vorst (Editors), Mathematical Finance-Bachelier Congress 2000 (2001)
M. Gundlach, F. Lehrbass (Editors), CreditRisk$^+$ in the Banking Industry (2004)
B.P. Kellerhals, Asset Pricing (2004)
Y.-K. Kwok, Mathematical Models of Financial Derivatives (1998)
M. Külpmann, Irrational Exuberance Reconsidered (2004)
P. Malliavin and A. Thalmaier, Stochastic Calculus of Variations in Mathematical Finance (2005)
A. Meucci, Risk and Asset Allocation (2005)
A. Pelsser, Efficient Methods for Valuing Interest Rate Derivatives (2000)
J.-L. Prigent, Weak Convergence of Financial Markets (2003)
B. Schmid, Credit Risk Pricing Models (2004)
S.E. Shreve, Stochastic Calculus for Finance I (2004)
S.E. Shreve, Stochastic Calculus for Finance II (2004)
M. Yor, Exponential Functionals of Brownian Motion and Related Processes (2001)
R. Zagst, Interest-Rate Management (2002)
Y.-L. Zhu, X. Wu, I.-L. Chern, Derivative Securities and Difference Methods (2004)
A. Ziegler, Incomplete Information and Heterogeneous Beliefs in Continuous-time Finance (2003)
A. Ziegler, A Game Theory Analysis of Options (2004)

John van der Hoek and Robert J. Elliott

Binomial Models in Finance

With 3 Figures and 25 Tables

John van der Hoek
Discipline of Applied Mathematics
University of Adelaide
Adelaide S.A. 5005 Australia
e-mail: john.vanderhoek@adelaide.edu.au

Robert J. Elliott
Haskayne School of Business
Scurfield Hall
University of Calgary
2500 University Drive NW
Calgary, Alberta, Canada T2N 1N4
e-mail:relliott@ucalgary.ca

Mathematics Subject Classification (2000): 91B28, 60H30

Library of Congress Control Number: 2005934996

ISBN-10 0-387-25898-1
ISBN-13 978-0-387-25898-0

Printed on acid-free paper.

© 2006 Springer Science+Business Media, Inc.
All rights reserved. This work may not be translated or copied in whole or in part without the written permission of the publisher (Springer Science+Business Media, Inc., 233 Spring Street, New York, NY 10013, USA), except for brief excerpts in connection with reviews or scholarly analysis. Use in connection with any form of information storage and retrieval, electronic adaptation, computer software, or by similar or dissimilar methodology now known or hereafter developed is forbidden.
The use in this publication of trade names, trademarks, service marks, and similar terms, even if they are not identified as such, is not to be taken as an expression of opinion as to whether or not they are subject to proprietary rights.

Printed in the United States of America.   (MVY)

9 8 7 6 5 4 3 2 1

springeronline.com

Acknowledgements

The authors wish to thank the Social Sciences and Humanities Research Council of Canada for its support. Robert Elliott gratefully thanks RBC Financial Group for supporting his professorship. John van der Hoek thanks the Haskayne Business School for their hospitality during visits to the University of Calgary to discuss the contents of this book. Similarly Robert Elliott wishes to thank the University of Adelaide. Both authors wish to thank various students who have provided comments and feedback when this material was taught in Adelaide, Calgary and St John's. The authors' thanks are also due to Andrew Royal for help with typing and formatting.

Preface

This book describes the modelling of prices of financial assets in a simple discrete time, discrete state, binomial framework. By avoiding the mathematical technicalities of continuous time finance we hope we have made the material accessible to a wide audience. Some of the developments and formulae appear here for the first time in book form.

We hope our book will appeal to various audiences. These include MBA students, upper level undergraduate students, beginning doctoral students, quantitative analysts at a basic level and senior executives who seek material on new developments in finance at an accessible level.

The basic building block in our book is the one-step binomial model where a known price today can take one of two possible values at a future time, which might, for example, be tomorrow, or next month, or next year. In this simple situation "risk neutral pricing" can be defined and the model can be applied to price forward contracts, exchange rate contracts and interest rate derivatives. In a few places we discuss multinomial models to explain the notions of incomplete markets and how pricing can be viewed in such a context, where unique prices are no longer available.

The simple one-period framework can then be extended to multi-period models. The Cox-Ross-Rubinstein approximation to the Black Scholes option pricing formula is an immediate consequence. American, barrier and exotic options can all be discussed and priced using binomial models. More precise modelling issues such as implied volatility trees and implied binomial trees are treated, as well as interest rate models like those due to Ho and Lee; and Black, Derman and Toy.

The book closes with a novel discussion of real options. In that chapter we present some new ideas for pricing options on non-tradeable assets where the standard methods from financial options no longer apply. These methods provide an integration of financial and actuarial pricing techniques.

Practical applications of the ideas and problems can be implemented using a simple spreadsheet program such as Excel. Many practical suggestions for implementing and calibrating the models discussed appear here for the first time in book form.

Contents

1 **Introduction** 1
   1.1 No Arbitrage and Its Consequences 1
   1.2 Exercises 11

2 **The Binomial Model for Stock Options** 13
   2.1 The Basic Model 13
   2.2 Why Is π Called a Risk Neutral Probability? 21
   2.3 More on Arbitrage 24
   2.4 The Model of Cox-Ross-Rubinstein 25
   2.5 Call-Put Parity Formula 27
   2.6 Non Arbitrage Inequalities 29
   2.7 Exercises 34

3 **The Binomial Model for Other Contracts** 41
   3.1 Forward Contracts 41
   3.2 Contingent Premium Options 43
   3.3 Exchange Rates 45
   3.4 Interest Rate Derivatives 55
   3.5 Exercises 61

4 **Multiperiod Binomial Models** 65
   4.1 The Labelling of the Nodes 65
   4.2 The Labelling of the Processes 65
   4.3 Generalized Quantities 66

	4.4	Generalized Backward Induction Pricing Formula 67
	4.5	Pricing European Style Contingent Claims 68
	4.6	The CRR Multiperiod Model 68
	4.7	Jamshidian's Forward Induction Formula 69
	4.8	Application to CRR Model 71
	4.9	The CRR Option Pricing Formula 73
	4.10	Discussion of the CRR Formula 75
	4.11	Exercises 78
5	**Hedging** 81
	5.1	Hedging 81
	5.2	Exercises 88
6	**Forward and Futures Contracts** 89
	6.1	The Forward Contract 89
	6.2	The Futures Contract 90
	6.3	Exercises 96
7	**American and Exotic Option Pricing** 97
	7.1	American Style Options 97
	7.2	Barrier Options 99
	7.3	Examples of the Application of Barrier Options 102
	7.4	Exercises 106
8	**Path-Dependent Options** 109
	8.1	Notation for Non-Recombing Trees 109
	8.2	Asian Options 110
	8.3	Floating Strike Options 112
	8.4	Lookback Options 113
	8.5	More on Average Rate Options 114
	8.6	Exercises 118

Contents XI

9 The Greeks 121
    9.1 The Delta (Δ) of an Option 121
    9.2 The Gamma (Γ) of an Option 123
    9.3 The Theta (Θ) of an Option 124
    9.4 The Vega (κ) of an Option 125
    9.5 The Rho (ρ) of an Option 125
    9.6 Exercises 126

10 Dividends 127
    10.1 Some Basic Results about Forwards 128
    10.2 Dividends as Percentage of Spot Price 129
    10.3 Binomial Trees with Known Dollar Dividends 132
    10.4 Exercises 134

11 Implied Volatility Trees 135
    11.1 The Recursive Calculation 136
    11.2 The Inputs V^{put} and V^{call} 138
    11.3 A Simple Smile Example 141
    11.4 In General 144
    11.5 The Barle and Cakici Approach 145
    11.6 Exercises 149

12 Implied Binomial Trees 153
    12.1 The Inputs 153
    12.2 Time T Risk-Neutral Probabilities 154
    12.3 Constructing the Binomial Tree 155
    12.4 A Basic Theorem and Applications 158
    12.5 Choosing Time T Data 161
    12.6 Some Proofs and Discussion 164
    12.7 Jackwerth's Extension 168
    12.8 Exercises 170

13 Interest Rate Models 171
13.1 $P(0,T)$ from Treasury Data 172
13.2 $P(0,T)$ from Bank Data 174
13.3 The Ho and Lee Model 184
13.4 The Pedersen, Shiu and Thorlacius Model 189
13.5 The Morgan and Neave Model 191
13.6 The Black, Derman and Toy Model 193
13.7 Defaultable Bonds 205
13.8 Exercises 205

14 Real Options 209
14.1 Examples 210
14.2 Options on Non-Tradeable Assets 214
14.3 Correlation with Tradeable Assets 229
14.4 Approximate Methods 233
14.5 Exercises 235

A The Binomial Distribution 237
A.1 Bernoulli Random Variables 237
A.2 Bernoulli Trials 239
A.3 Binomial Distribution 239
A.4 Central Limit Theorem (CLT) 243
A.5 Berry-Esséen Theorem 245
A.6 Complementary Binomials and Normals 246
A.7 CRR and the Black and Scholes Formula 247

B An Application of Linear Programming 249
B.1 Incomplete Markets 250
B.2 Solutions to Incomplete Markets 251
B.3 The Duality Theorem of Linear Programming 253
B.4 The First Fundamental Theorem of Finance 257
B.5 The Duality Theorem 261
B.6 The Second Fundamental Theorem of Finance 264
B.7 Transaction Costs 266

Contents XIII

- **C Volatility Estimation** 269
  - C.1 Historical Volatility Estimation 270
  - C.2 Implied Volatility Estimation 272
  - C.3 Exercises 278

- **D Existence of a Solution** 279
  - D.1 Farkas' Lemma 279
  - D.2 An Application to the Problem 281

- **E Some Generalizations** 285
  - E.1 Preliminary Observations 285
  - E.2 Solution to System in van der Hoek's Method 287
  - E.3 Exercises 288

- **F Yield Curves and Splines** 289
  - F.1 An Alternative representation of Function (F.1) 290
  - F.2 Imposing Smoothness 291
  - F.3 Unknown Coefficients 291
  - F.4 Observations 292
  - F.5 Determination of Unknown Coefficients 293
  - F.6 Forward Interest Rates 295
  - F.7 Yield Curve 296
  - F.8 Other Issues 296

References 297

Index 301

1

Introduction

1.1 No Arbitrage and Its Consequences

The prices we shall model will include prices of underlying assets and prices of derivative assets (sometimes called **contingent claims**).

Underlying assets include commodities, (oil, gas, gold, wheat,...), stocks, currencies, bonds and so on. **Derivative assets** are financial investments (or contracts) whose prices depend on other underlying assets.

Given a model for the underlying asset prices we shall deduce prices for derivative assets. We shall model prices in various markets, equities (stocks), foreign exchange (FX). More advanced topics we shall discuss include incomplete markets, transaction costs, credit risk, default risk and real options.

As Newtonian mechanics is based on axioms known as Newton's laws of motion, derivative pricing is usually based on the axiom that there is **no arbitrage opportunity**, or as it is sometimes colloquially expressed, **no free lunch**.

There is only one **current state** of the world, which is known to us. However, a future state at time T is unknown; it may be one of many possible states.

An arbitrage opportunity is a little more complicated than saying we can start now with nothing and end up with a positive amount. This would, presumably, mean we end up with a positive amount in all possible states at the future time. In Chapter 2, we shall meet two forms of arbitrage opportunities. For the moment we shall discuss one of these which we shall later refer to as a "type two arbitrage opportunity".

Definition 1.1 (Arbitrage Opportunity). *More precisely, an arbitrage opportunity is an asset (or a portfolio of assets) whose value today is zero and whose value in all possible states at the future time is never negative, but in some state at the future time the asset has a strictly positive value.*

In notation, suppose $W(0)$ is the value of an asset (or portfolio) today and $W(T,\omega)$ is its value at the future time T when the state of the world is ω. Then an arbitrage opportunity is some financial asset W such that

$$W(0) = 0$$
$$W(T,\omega) \geq 0 \text{ for all states } \omega$$
$$\text{and } W(T,\omega) > 0 \text{ for some state } \omega$$

Our fundamental axiom is then:

Axiom 1 *There are no such arbitrage opportunities.*

A consequence of this axiom is the following basic result:

Theorem 1.2 (Law of One Price). *Suppose there are two assets A and B with prices at time 0 $P_0(A) \geq 0, P_0(B) \geq 0$. Supposing at some time $T \geq 0$ the prices of A and B are equal in all states of the world:*

$$P_T(A) = P_T(B).$$

Then

$$P_0(A) = P_0(B).$$

Proof. We shall show that otherwise there exists an arbitrage. Without loss of generality, suppose that $P_0(A) > P_0(B)$. We construct the following portfolio at time 0. Starting with $0:

We borrow and sell A. This realizes	$P_0(A)$
We buy B; this costs	$-P_0(B)$

So this gives a positive amount $P_0(A) - P_0(B)$, which we can keep, or even invest. Note this strategy requires no initial investment. At time T we clear our books by:

Buying and returning A. This costs	$-P_T(A)$
Selling B, giving	$P_T(B)$
Net cost is	$0

However, we still have the positive amount $P_0(A) - P_0(B)$, and so we have exhibited an arbitrage opportunity. Our axiom rules these out, so we must have $P_0(A) = P_0(B)$. □

1.1 No Arbitrage and Its Consequences

In this proof we have assumed there are no transaction costs in carrying out the trades required, and that the assets involved can be bought and sold at any time at will. The imposition and relaxing of such assumptions are part of financial modelling.

We shall use the **one price** result to determine a rational price for derivative assets.

As our first example of a derivative contract, let us introduce a **forward contract**. A forward contract is an agreement (a contract) to buy or sell a specified quantity of some underlying asset at a **specified price**, with delivery at a **specified time and place**.

The buyer in any contract is said to take the **long** position. The seller in any contract is said to take the **short** position.

The **specified delivery price** is agreed upon by the two parties at the time the contract is made. It is such that the (initial) cost to both parties in the contract is 0.

Most banks have a **forward desk**. It will give quotes on, say, the exchange rate between the Canadian dollar and U.S. dollar.

Example 1.3. U.S.$/C$

SPOT	0.7540
60 DAY FORWARD	0.7510
90 DAY FORWARD	0.7495
180 DAY FORWARD	0.7485

Forward contracts can be used for **hedging** and **speculation**.

Hedging

Suppose a U.S. company knows it must pay a C$1 million in 90 days' time. At no cost it can enter into a forward contract with the bank to pay

$$U.S.\$749,500.$$

This amount is agreed upon today and fixed. Similarly, if the U.S. company knows it will receive C$1 million in 90 days, it can enter into a short forward contract with the bank to sell C$1 million in 90 days for

$$U.S.\$749,500.$$

Speculation

An investor who thinks the C$ will increase against the US.$ would take a long position in the forward contract agreeing to buy C$1 million for

U.S.$749,500

in 90 days' time.

Suppose the U.S.$/C$ exchange rate in 90 days is, in fact, 0.7595. Then the investor makes a profit of

$$10^6 \times (0.7595 - 0.7495) = \text{U.S.\$10,000}.$$

Of course, forward contracts are **binding** and if, in fact, the U.S.$/C$ exchange rate in 90 days is 0.7395 then the investor must still buy the C$1 million for U.S.$749,500.

However, the market price of C$1 million is only U.S.$739,500, and so the investor realizes a loss of U.S.$10,000.

Let us write S_0 for the price of the underlying asset today and S_T for the price of the asset at time T. Write K for the agreed price. The profit for a long position is then $S_T - K$, a diagram of which is shown in Figure 1.1.

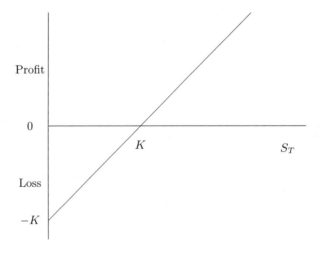

Fig. 1.1. The payoff of a long forward contract.

The profit for a short position in a forward contract is $K - S_T$, a diagram of which is shown in Figure 1.2.

Either the long or short party will lose on a forward contract. This problem is managed by **futures contracts** in which the difference between the agreed

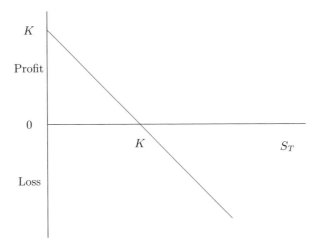

Fig. 1.2. The payoff of a short forward contract.

price and the spot price is adjusted daily. Futures contracts will be discussed in a later chapter.

In contrast to forward contracts which are binding, we wish to introduce options.

Definition 1.4 (Options). *A **call option** is the right, but not the obligation, to buy some asset for a specified price on or before a certain date.*

*A **put option** is the right, but not the obligation, to sell some asset for a specified price on or before a certain date.*

Remark 1.5. Unlike the forward contract, an option is not binding. The holder is not **obliged** to buy or sell. This, of course, gives rise to the term 'option'.

Call and put options can be **European** or **American**. This has nothing to do with the geographical location. European options can be exercised only on a certain date, the **exercise date**. American options can be exercised any time between now and a future date T (the **expiration time**). T may be $+\infty$, in which case the option is called **perpetual**.

To be specific we shall consider how call and put options are reported in the financial press.

Example 1.6. Consider Table 1.1 for Listed Option Quotations in the *Wall Street Journal* of July 23, 2003. These are examples of options written on

1 Introduction

Table 1.1. Listed Option Quotations

OPTION	STRIKE	EXP	-CALL-		-PUT-	
			VOL	LAST	VOL	LAST
AMR	11.0	Aug	3235	0.60	422	0.90
AOL TW	15.0	Aug	8152	2.00	494	0.20
16.85	16.0	Aug	3317	1.20	721	0.45
16.85	17.5	Aug	6580	0.45	1390	1.20

common stock or shares. Consider the table and the entries for AOL TW (America Online/Time Warner). The entry of $16.85 under AOL TW gives the closing price on Tuesday, July 22, 2003, of AOL TW stock. Note that for the first entry AMR (American Airlines), only one option and put was traded. The AMR entry is given on one line and its closing price of $10.70 is omitted.

The second column gives the strike, or exercise, price of the option. The first option for AOL has a strike price of $15, the line below refers to a strike of $16 and the third line for AOL refers to a strike of $17.50.

The third column refers to the expiry month. Stock options expire on the third Friday of their expiry month.

Of the last four columns, the first two refer to call options and the final two to put options. The VOL entry gives the number of CALL or PUT options sold. The LAST entry gives the closing price of the option. For example, the closing price of an AOL August call with strike price $15 was $2; the closing price of an AOL August put with strike price $15 was $0.20.

Of course, the price of a stock may vary throughout a day. What is taken as the representative price of a stock for a particular day is a matter of choice. This book will not deal with **intraday modelling of price movements**. However, **Reuter Screens**, and the like, present data on prices on an almost continuous basis.

We shall shortly write down **models** for the evolution of stock prices. S will be the **underlying process** for the options here. S will just be called the **underlying**.

To be definite let us write

$$S = \{S(t) \mid t \geq 0\}$$

for the price process of this stock (the **stock price process**).

Call Options

In order to specify a call option contract, we need three things:

1. an **expiry date**, T (also called the **maturity date**);
2. a **strike price**, K (or also called the **exercise price**);
3. a **style** (European, American or even Bermudan, etc).

Let us discuss the AUG 2003 AOL Call options, for example the AOL/AUG/15.00/CALL. This means that the strike price is $15.00. We will write $K = \$15.00$. The expiry date is August 2003. As we are dealing with an **exchange traded option** (ETO) on the **New York Stock Exchange** (NYSE), this will mean: 10:59 pm Eastern Time on the Saturday following the third Friday of the expiration month. An investor holding the option has until 4:30 pm on that Friday to instruct his or her broker to exercise the option. The broker then has until 10:59 pm the following day to complete the paperwork effecting that transaction. In 2003, the August contract expired on August 15, the third Friday of August.

Time is measured in years or fractions of years. In 2003, there were 24 calendar days from July 22 to expiry, (22 July to 15 August); this is $\frac{24}{365} = 0.06575$ years. This is the way we shall calculate time. Another system is to use trading days, of which there are about 250 in a year. As there are 18 trading days from 22 July until 15 August, we would get $\frac{18}{250} = 0.072$ years. There is another convention that there are 360 days in a year. This is common in the United States.

The holder of a call option owns a contract which gives him/her the (legal) right (but not the obligation) to buy the stock at any time up to and including the expiry date for the strike (or exercise) price.

This is an example of an American (style) call option. An **American style option** is one that can be exercised at any time up to and including the expiry date. On the other hand, as we have noted, a **European style option** is one that can be exercised only on the expiry date. **Mid-Atlantic** or **Bermuda style options** are ones that are halfway between American and European style options. For example we could require that the option only be exercised on a Thursday.

Usually, one enters a call option contract by the payment of a fee, which is called the **option price**, the **call price** or the **call premium**. However, it is possible to vary the style of payment—pay along the way until expiry, pay at expiry and so on. It is one of the goals of this book to determine the rational price, or premium, for a call option. This leads us to the area of **option pricing**.

If you are **long in an American call** (that is, you own the call option), then at any time prior to the expiry date, you can do one of three things:

1. sell the call to someone else;
2. exercise the call option—that is, purchase the underlying stock for the agreed strike price K;
3. do nothing.

If you own a European style call option, only choices 1. and 3. are possible as the option can be exercised only at the expiry date.

In this book we shall provide **option pricing formulas**, but the market also provides option prices, (determined in the exchange by an auction process). Hopefully, the theoretical and the market valuations will agree, at least to a good approximation.

Some Basic Notions

For most financial assets there is a selling (asking) price and a buying (bid) price. Why is the selling (asking) price always greater than the buying (bid) price? If the bid price were greater than or equal than the asking price, **the market would clear** all mutually desirable trades until the asking price were strictly greater than the bid price.

We shall usually make the simplifying assumption that there is one price for both sellers and buyers at any one time. This also means that we shall ignore **transaction costs**. This is one of the reasons for bid-ask spreads. At a later stage we shall address the issue of bid-ask spreads.

What is the value of the call option at expiry? Let T be the expiry time. Then for $0 \leq t \leq T$, let $C(t)$ be the value of the call option at time t. We claim that

$$C(T) = \max\{0, S(T) - K\} = (S(T) - K)^+ \tag{1.1}$$

where for any number a, $a^+ = \max\{0, a\}$. To see this we can consider three cases: (1) $S(T) > K$; (2) $S(T) = K$; (3) $S(T) < K$. In the first situation, we could exercise the option, purchasing the stock for \$K and then selling the stock at the market price \$$S(T)$ to realize a profit of \$$(S(T) - K)$. This, of course, assumes no transaction costs that would reduce this profit. In the second and third cases we would not exercise the option, but let it lapse, as it would be cheaper to buy the stock at the market price.

Let us also note that for an American style call option

$$C_A(t) \geq (S(t) - K)^+ \geq 0 \tag{1.2}$$

where we write $C_A(t)$ for the American option price.

The reason for (1.2) is clear: If we exercise the option and $S(t) > K$ then the exercise value is $(S(t) - K)^+$; if we do not exercise, this may be because the value of holding the option is greater than the present exercise value.

1.1 No Arbitrage and Its Consequences

The value $C(T)$ at expiry is uncertain when viewed from the present, because $S(T)$ is uncertain. However, we shall determine $C(0)$ and $C(t)$ for $0 \leq t \leq T$.

A call option is an example of a **derivative** (or **derived asset**) because its value is dependent on (is contingent on) the value of an underlying asset (or price process) in this case a stock price process S. So **derivative** equals **derived asset** equals **contingent claim**. An option is called an **asset** as it is something that can be bought and sold.

Why is there a market for call options? This is an important question as there may be no potential buyers and sellers. This question, of course, applies to any asset. For this discussion let us focus on the simpler European call option.

Let us first note that there are basically three types of players in financial markets:

1. **speculators** (or risk takers, investors, and so on);
2. **hedgers** (or risk avertors);
3. **arbitrageurs** (looking for mispriced assets).

For the meantime let us focus on 1. and 2. When we have discussed derivative pricing, we shall discuss possible strategies (arbitrage opportunities) when mispricing occurs. The existence of arbitrageurs keeps prices at **fair values**.

Later on we shall consider other financial products from the point of view of 1., 2. or 3.

In each of 1. and 2., the market players will **take a view about the future**. For example, 1. may assume that prices of a stock will go up. Such a player is said to be **bullish** (as opposed to being **bearish**). Once a view has been taken, then a financial product can be used to **profit from this view if it is realized**.

Buying a call option (taking a call, being long in a call). Suppose S refers to AOL stock. Here are two strategies that give rise to the purchase of call options.

1. **Leverage** is a speculator's strategy. At present (22 July 2003, say), $S(0) = \$16.85$, and we suppose that on the 15 August 2003 (the expiry date of the AUG2003 option), that $S(T) = \$18.00$. Suppose that you have $1685 at your disposal, a convenient amount.

    You could buy 100 shares @ $16.85, and if your view is realized on 15 August 2003, you could make a profit of $100 \times (\$18.00 - \$16.85) = \$115$ which is a 6.82% profit ($\frac{115}{1685} \times 100 = 6.82\%$). Suppose now that the view was not realized and that the stock price fell to $15.00. Then you would suffer a loss of $\$185 = 100 \times (\$16.85 - \$15.00)$ or 10.98% in percentage terms.

An alternative to buying stock is to obtain leverage using options. Instead, consider buying 1000 AOL/AUG/16.00/CALL options at $1.20 each (a convenient approximation). We shall ignore transaction costs, and the question of whether there are 1000 options available to be purchased. If the view is realized on 15 August 2003, then you have $1000 \times (18.00 - 16.00) = \$2000$, which gives a profit of $\$(2000 - 1200) = \$800$ (equal to 66.67% in percentage terms). If your view was not realized and the stock price fell to $15.00, then you would have $0, and so you have a 100% loss. Therefore, options **magnify** or **leverage** profits if views are realized, but on the downside you can lose all you put down (**but no more**).

With some **exotic options** it is possible to obtain higher leverage. However, we would have to purchase these products **over the counter** (OTC) rather than through an exchange. Note that speculators are using **out of the money** call options to obtain leverage. Also, note that on 22 July 2003 in-the-money calls with $K = 15.00$ or 16.00 had volumes 8152 and 3317 respectively; out-of-the-money calls with $K = 17.50$ had a volume of 6580.

2. **Hedging** is a risk avertor strategy. A risk avertor will buy options now to lock in a fixed future price, at which he has the option to buy a share, no matter what actually happens to the stock price. Suppose that on 22 July 2003 you decided that you wished to buy AOL shares on 15 August 2003 for $17.00, but you are worried that the share price may rise to $18.00. You could then buy AOL/AUG/17.00/CALL options. If the fear were realized, you would only need to pay $17.00 for each share. Of course, if the share price fell to $15.00, then you would not exercise the option but buy the shares in the market for this lower price. The payment of the premiums for these call options can be regarded as an **insurance payment** against the possible rise in price of the stock price. This strategy usually uses ATM call options, that is, at the money call options with $K = S(0)$.

Selling a call option (**writing a call, being short in a call**). "Selling calls" is also called "writing calls" as the seller of a call option writes the contract. The opposite of a **writer** is a **taker** (the buyer). There are several strategies that give rise to writing call options.

1. **Income generation**. If you own shares, you can write call options on these stocks to generate extra income from holding the shares by way of collecting premiums. It is like an extra **dividend** on the shares. If you do this, you must **be prepared to sell the shares**, or **be able to sell the shares**, if the call options are exercised against you. Most call writers who adopt this strategy actually hope that the calls will **not be exercised**. In order to have some guarantee of this the calls should be **out of the money** call options. This strategy is often called the **buy and write** strategy, and is widely used by investment houses.

This strategy uses the **covered call**, whereas if you write call options on stock that you do not own, you are said to be writing a **naked call**. This latter strategy is used by some speculators. However, it is dangerous in that if the call is exercised, the writer of the call will have to buy the stock at market price and deliver it at a possibly lower price, so incurring a possible loss.

2. **Insurance.** If you have the view that share prices will fall, you may be interested in selling call options to generate income that will compensate you for the falling share prices. However, there is only limited protection from this strategy. You would use out-of-the-money call options and be protected from a loss down to $S(0) - C(0)$, which could be rather limited. Of course, here **put options** are a more natural instrument for insurance. Buying a put with a strike of $\$K$ ensures one can always sell the underlying for $\$K$. This provides a minimum value for one's holdings in the underlying.

In Summary

Let us note in summary that **both buyers and sellers of calls are mainly interested in out-of-the-money calls**. This is just as well, for if the buyers wanted in-the-money call options and the sellers only provided out-of-the-money call options, there would not be a market!

We could have carried out a similar discussion for put options. These are contracts structured just as calls, but the holder of a put has the right but not the obligation to **sell** the stock at the strike price at (or before) the expiry date. Of course, there are European style puts, American style puts, and Bermudan puts, and so on.

Remark 1.7. Because most traded options are of American style, and because many of these are out-of-the-money options, they are rarely exercised early.

1.2 Exercises

Exercise 1.8. We have provided motivation for the buying and selling of call options and we have noted that, in general, the needs of buyers and sellers can be matched. Carry out a similar discussion for put options.

2

The Binomial Model for Stock Options

2.1 The Basic Model

We now discuss a simple **one-step binomial model** in which we can determine the rational price today for a call option. In this model we have two times, which we will call $t = 0$ and $t = 1$ for convenience. The time $t = 0$ denotes the present time and $t = 1$ denotes some future time. Viewed from $t = 0$, there are two states of the world at $t = 1$. For convenience they will be called the **upstate** (written ↑) and the **downstate** (written ↓). There is **no special meaning** to be attached to these states. It does not **necessarily** mean that a stock price has a low price in the downstate and a higher value in the upstate, although this will sometimes be the case. The term **binomial** is used because there are **two** states at $t = 1$.

In our model there are two **tradeable assets**; eventually there will be other derived assets:

1. **a risky asset** (e.g. a stock);
2. **a riskless asset**.

By a **tradeable asset** we shall mean an asset that can be bought or sold on demand at any time in any quantity. They are the typical assets used in the construction of portfolios. In Chapter 14 on real options we shall note some problems with this concept.

We assume for each asset that its buying and selling prices are equal.

The risky asset.

At $t = 0$, the risky asset S will have the known value $S(0)$ (often non-negative).

At $t = 1$, the risky asset has two distinct possible values (hence its value is uncertain or risky), which we will call $S(1,↑)$ and $S(1,↓)$. We simply require that $S(1,\uparrow) \neq S(1,\downarrow)$, but without loss of generality (wlog), we may assume that $S(1,\uparrow) > S(1,\downarrow)$.

The riskless asset

At $t = 0$, the riskless asset B will have value $B(0) = 1$.

At $t = 1$, the riskless asset has the same value (hence riskless) in both states at $t = 1$, so we write $B(1,\uparrow) = B(1,\downarrow) \equiv R = 1 + r$. Usually $R \geq 1$ and so $r \geq 0$, which we can call **interest**, is non-negative. It represents the amount earned on \$1.

It is easy to show that if $S(1,\uparrow) = S(1,\downarrow)$ there is an arbitrage, unless $S(1,\uparrow) = S(1,\downarrow) = (1+r)S(0)$.

We also assume that

$$S(1,\downarrow) < RS(0) < S(1,\uparrow). \tag{2.1}$$

We shall see the importance of inequality (2.1) below.

Example 2.1. Here $S(0) = 5$, $S(1,\uparrow) = \frac{20}{3}$ and $S(1,\downarrow) = \frac{40}{9}$. $B(0) = 1$ and $B(1,\uparrow) = B(1,\downarrow) = R = \frac{10}{9}$. So $r = \frac{1}{9}$ and (2.1) clearly holds.

Suppose $X(1)$ is any claim that will be paid at time $t = 1$. In our model $X(1)$ can take one of two values: $X(1,\uparrow)$ or $X(1,\downarrow)$. We shall determine $X(0)$, the premium or price of X at time $t = 0$.

Often the values of $X(1)$ are uncertain because $X(1) = f(S(1))$ (a function of S) and $S(1)$ is uncertain. As X is an asset whose value depends on S, it is a **derived asset** written on S, or a derivative on S. X is also called a **derivative** or a **contingent claim**.

Example 2.2. When we write $X(1) = [S(1) - K]^+$ we mean

$$X(1,\uparrow) = [S(1,\uparrow) - K]^+$$
$$X(1,\downarrow) = [S(1,\downarrow) - K]^+.$$

Assuming we have a model for S, we can find $X(0)$ in terms of this information. This could be called **relative pricing**. It presents a different methodology than, (though often equivalent to) what the economists call **equilibrium pricing**, for example.

There are two steps to relative pricing.

Step 1

Find H_0 and H_1 so that

2.1 The Basic Model

$$X(1) = H_0 B(1) + H_1 S(1). \tag{2.2}$$

Both sides here are random quantities and (2.2) means

$$X(1,\uparrow) = H_0 R + H_1 S(1,\uparrow) \tag{2.3a}$$
$$X(1,\downarrow) = H_0 R + H_1 S(1,\downarrow). \tag{2.3b}$$

The interpretation is as follows: H_0 represents the number of dollars held at $t = 0$, and H_1 the number of stocks held at $t = 0$. At $t = 1$, the level of holdings does not change, but the underlying assets do change in value to give $H_0 B(1) + H_1 S(1)$.

Solving (2.3a) and (2.3b) gives

$$H_1 = \frac{X(1,\uparrow) - X(1,\downarrow)}{S(1,\uparrow) - S(1,\downarrow)} \tag{2.4}$$

and

$$\begin{aligned} H_0 &= \frac{X(1,\uparrow) - H_1 S(1,\uparrow)}{R} \\ &= \frac{X(1,\uparrow) - \frac{X(1,\uparrow)-X(1,\downarrow)}{S(1,\uparrow)-S(1,\downarrow)} S(1,\uparrow)}{R} \\ &= \frac{S(1,\uparrow)X(1,\downarrow) - S(1,\downarrow)X(1,\uparrow)}{R\left[S(1,\uparrow) - S(1,\downarrow)\right]}. \end{aligned} \tag{2.5}$$

Note: It is rather crucial that $S(1,\uparrow) \neq S(1,\downarrow)$.

Example 2.3 (continuation of Example (2.1)). If $X(1,\uparrow) = 7$ and $X(1,\downarrow) = 2$, then equations (2.3a) and (2.3b) become

$$7 = H_0 \frac{10}{9} + H_1 \frac{20}{3}$$
$$2 = H_0 \frac{10}{9} + H_1 \frac{40}{9},$$

giving $H_0 = -7.2$ and $H_1 = 2.25$.

Remark 2.4. We should now take a little time to interpret the situation where H_0 or H_1 is negative.

In the previous example, $H_0 = -7.2$ means we borrowed 7.2 and $t = 0$ and we have a liability (a negative amount) of $H_0 R = -8$ at $t = 1$.

Suppose instead that $X(1,\uparrow) = 2$ and $X(1,\downarrow) = 7$, then $H_0 = 15.3$ and $H_1 = -2.25 < 0$. Now $H_1 = -2.25$ means we shorted (borrowed) 2.25 stocks at $t = 0$ and we have a liability at $t = 1$ as we must return the value of the stock at $t = 1$. This value will depend on whether we are in $\uparrow$ or $\downarrow$. By the way, we must also assume that we have a **divisible market**, which is one in which any (real) number of stocks can be bought and sold. If we think of stocks in lots of 1000 shares, then 2.25 is really 2250 shares. This is how we could interpret these "fractional shares".

Short sell means "borrow and sell what you do not own".

There are basically two ways of raising cash: Borrow money at interest (from a bank, say) or short sell an asset. In the former case, you must repay the loan with interest at a future date and in the second case, you must buy back the asset later and return it to its owner.

In an analogous way there are two ways of devolving yourself of cash. You can put money in a bank to earn interest, or you can buy an asset. In the former case you can remove the money later with any interest it has earned, and in the latter case you can sell the asset (at a profit or loss) at a future date.

Step 2

Using the **one price theorem**, which is a consequence of the **no arbitrage axiom**, we must have

$$X(0) = H_0 + H_1 S(0). \tag{2.6}$$

Remark 2.5. This equation is true because the claim X and the portfolio $H_0 B + H_1 S$ have the same value in both possible states of the world at $t = 1$. In this situation, $X(0)$ represents **outflow** of cash at $t = 0$. If $X(0) > 0$, then $X(0)$ represents the amount to be paid at $t = 0$ for the asset with payoff $X(1)$ at $t = 1$. If $X(0) < 0$, then $-X(0)$ represents an amount received at $t = 0$ for the asset with payoff $X(1)$ at $t = 1$.

We shall review for this call option why $X(0)$ must equal $H_0 + H_1 S_0$. First assume (if possible) that

$$X(0) < H_0 + H_1 S(0). \tag{2.7}$$

In fact let us use the numbers from the previous example. Thus (2.7) is

$$2.25 S(0) - 7.2 - X(0) > 0 \tag{2.8}$$

We now perform the following trades at t=0.

Short sell 2.25 shares of stock, put 7.2 in the bank, buy one asset.

2.1 The Basic Model

Equation (2.8) gives the strategy to adopt. If a quantity is a positive value of assets such as $2.25S(0)$, this suggests one should short sell the assets; if a quantity is a negative value of assets (that is, $-X(0)$), this suggests one should purchase the assets. A positive number alone indicates a borrowing and a negative number, -7.2, an investment of cash in a bank.

In fact
$$2.25S(0) - (7.2 + X(0)) > 0$$
where $2.25S(0)$ is income, $7.2+X(0)$ payouts. Note that because this difference is positive you have a profit from this trading at $t = 0$. Put this profit in your pocket—and do not touch it (at least for the time being).

Note the following: You did not need any of your own money to carry out this trade. The short sale of the borrowed stock was enough to finance the investment of 7.2 and the purchase of X for $X(0)$, and there was money left over.

The consequence at t=1.

There are two cases:

In ↑

Sell X for $X(1,\uparrow) = 7$, remove the money from the bank with interest $7.2R = 8$. This results in 15 (dollars), which can be used to fund the repurchase (and return) of the $2.25S(1,\uparrow) = 15$. There are no further liabilities. Thus, there are **no unfunded liabilities at t=1**.

In ↓

Sell X for $X(1,\downarrow) = 2$, remove the money from the bank with interest $7.2R = 8$. This results in 10 (dollars), which can be used to fund the repurchase (and return) of the $2.25S(1,\downarrow) = 10$. There are again no further liabilities. Thus again there are **no unfunded liabilities at $t = 1$**.

In summary, we have made a profit at $t = 0$ and have no unfunded liabilities at $t = 1$. This is making money by taking no risks—by not using your own money. This is an example of an arbitrage opportunity which our fundamental axiom rules out. In efficient markets one assumes that arbitrage opportunities do not exist, and so we have a contradiction to (2.8). In practice, arbitrage opportunities may exist for brief moments, but, due to the presence of arbitrageurs, the markets quickly adjust prices to eliminate these arbitrage opportunities. At least that is the theory.

After this discussion we see that (2.7) cannot hold (at least not in the example, but also more generally). Therefore,

$$X(0) \geq H_0 + H_1 S(0).$$

Assume now, if possible, that

$$X(0) > H_0 + H_1 S(0). \qquad (2.9)$$

In the example, this would mean

$$X(0) + 7.2 - 2.25 S(0) > 0. \qquad (2.10)$$

We now perform the following trades at t=0.

Short sell the asset, borrow 7.2 and buy 2.25 stock.

This yields a positive profit at $t = 0$ which is placed deep in your pocket until after $t = 1$. In other words raising funds from the short sale and borrowings is more than enough to cover the cost of 2.25 shares.

The consequence at t=1.

There are two cases:

In $\uparrow$

Sell the shares for $2.25 S(1, \uparrow) = 15.00$, repay the loan with interest $7.2R = 8$, purchase the asset for 7 and return to the (rightful) owner. Everything balances out. Thus, there are **no unfunded liabilities at t=1**.

In $\downarrow$

Sell the shares for $2.25 S(1, \downarrow) = 10.00$, repay the loan with interest $7.2R = 8$, purchase the asset for 2 and return to the (rightful) owner. Everything balances out. Thus, there are **no unfunded liabilities at t=1**.

In summary, we have again made a profit at $t = 0$ and have no unfunded liabilities at $t = 1$. This is again an arbitrage opportunity. Therefore, (2.9) is false as well. We then conclude the result claimed in (2.6) must hold.

Let us now substitute (2.4) and (2.5) into (2.6). Then

$$\begin{aligned}
X(0) &= H_0 + H_1 S(0) \\
&= \left[\frac{S(1,\uparrow)X(1,\downarrow) - S(1,\downarrow)X(1,\uparrow)}{R\left(S(1,\uparrow) - S(1,\downarrow)\right)} \right] + \left[\frac{X(1,\uparrow) - X(1,\downarrow)}{S(1,\uparrow) - S(1,\downarrow)} \right] S(0) \\
&= \frac{X(1,\uparrow)\left[RS(0) - S(1,\downarrow)\right] + X(1,\downarrow)\left[S(1,\uparrow) - RS(0)\right]}{R\left[S(1,\uparrow) - S(1,\downarrow)\right]} \\
&= \frac{1}{R}\left[\pi X(1,\uparrow) + (1-\pi)X(1,\downarrow)\right]
\end{aligned}$$

where

$$\pi = \frac{RS(0) - S(1,\downarrow)}{S(1,\uparrow) - S(1,\downarrow)} > 0 \qquad (2.11)$$

$$1 - \pi = \frac{S(1,\uparrow) - RS(0)}{S(1,\uparrow) - S(1,\downarrow)} > 0.$$

Here $0 < \pi < 1$ follows from the assumption of the model (2.1).
Therefore,

$$X(0) = \frac{\pi X(1,\uparrow) + (1-\pi)X(1,\downarrow)}{R}. \tag{2.12}$$

This is the **general pricing formula** for a contingent claim option in a one-step binomial model.

It was derived by using two ideas:

1. replicating portfolios (step 1);
2. there are no arbitrage opportunities (vital for the step 2 argument).

This method is called **relative pricing** because relative to the given inputs $S(0)$, $S(1,\uparrow)$, $S(1,\downarrow)$, $B(0)$, $B(1,\uparrow)$ and $B(1,\downarrow)$ we can price other assets. We simply calculate π as in (2.11) and then use (2.12). Let us note that even though S was thought of as being a stock, it could have stood for **any** risky asset at all.

The numbers π and $1 - \pi$ are called the **risk neutral probabilities** of states $\uparrow$ and $\downarrow$, respectively. We shall see why this name is used.

We can write (2.12) as

$$X(0) = \mathbf{E}^\pi \left[\frac{X(1)}{B(1)} \right], \tag{2.13}$$

which is the **risk neutral expectation** of $\frac{X(1)}{B(1)}$. It stands for

$$\pi \frac{X(1,\uparrow)}{B(1,\uparrow)} + (1-\pi)\frac{X(1,\downarrow)}{B(1,\downarrow)}.$$

This is the same as the right hand side of (2.12).

Remark 2.6. It can be shown that there is no arbitrage possible in our binomial model if and only if (iff) a formula of the type (2.13) holds with $0 < \pi < 1$.

Remark 2.7. The author that is credited with the first use of binomial option pricing is Sharpe in 1978 [70, pages 366–373]. He argues as follows: First select h so that
$$hS(1,\uparrow) - X(1,\uparrow) = hS(1,\downarrow) - X(1,\downarrow) .$$
Set this common value equal to

$$R(hS(0) - X(0)).$$

This again leads to equation (2.12).

In 1979 Rendleman and Bartter [63] gave a similar argument. First select α so that

$$S(1,\uparrow) + \alpha X(1,\uparrow) = S(1,\downarrow) + \alpha X(1,\downarrow)$$

and set this common value to

$$R(S(0) + \alpha X(0)).$$

This (normally) again leads to equation (2.12). We say this because a choice of α may not always exist. For the Sharpe approach, a choice of h can always be made.

Exercise 2.8. Verify the claims made in this remark.

Not all models that one could write down are arbitrage free.

Example 2.9 (Continuation of Example 2.1).
Simply make the change $S(1,\downarrow) = \frac{17}{3}$. Starting with nothing, choose $H_0 = -5$ (borrow 5 stocks), $H_1 = 1$ (buy one stock). Then $H_0 + H_1 S(0) = 0$. At $t = 1$, our position will be $X(1) \equiv -5R + S(1)$ (meaning sell the stock and repay the loan). This is $\frac{10}{9}$ in the upstate and $\frac{1}{9}$ in the down state. So with no start-up capital we have generated a profit (in both states) by simply trading. This is an arbitrage opportunity. Note that condition (2.1) is violated here.

Example 2.10 (On why $0 < \pi < 1$ should hold). As in equation (2.11)

$$\pi = \frac{RS(0) - S(1,\downarrow)}{S(1,\uparrow) - S(1,\downarrow)}.$$

We assumed in inequality (2.1) that $0 < S(1,\downarrow) < RS(0) < S(1,\uparrow)$. So, for example,

$$0 < RS(0) - S(1,\downarrow) < S(1,\uparrow) - S(1,\downarrow)$$

and the result that (2.1) implies that $0 < \pi < 1$ follows. If we choose X with $X(1,\uparrow) = 1$ and $X(1,\downarrow) = 0$, then $X(0) > 0$ to exclude arbitrage. Then (2.12) implies that $\pi > 0$. A similar argument using X with $X(1,\uparrow) = 0$ and $X(1,\downarrow) = 1$ leads to $1 - \pi > 0$. So the absence of arbitrage opportunities leads to $0 < \pi < 1$.

Notation

It is often useful to use the following notation when $\mathbf{x} = (x_1, x_2, \ldots, x_n) \in \mathbf{R}^n$:

1. $\mathbf{x} \geq 0$ if $x_i \geq 0$ for each $i = 1, 2, \ldots, n$.
2. $\mathbf{x} > 0$ if $\mathbf{x} \geq 0$ and $x_i > 0$ for at least one i.
3. $\mathbf{x} \gg 0$ if $x_i > 0$ for each $i = 1, 2, \ldots, n$.

2.2 Why Is π Called a Risk Neutral Probability?

This discussion will take place within the **one-step binomial asset pricing model**.

Some of the steps here will be left to the reader as exercises.

For any $0 \leq p \leq 1$, let $\mathbf{E}^p[X(1)]$ be defined by

$$\mathbf{E}^p[X(1)] = pX(1,\uparrow) + (1-p)X(1,\downarrow). \tag{2.14}$$

Here p could represent a (subjective) probability (viewed from $t=0$) that the upstate ($\uparrow$) will occur at $t=1$. Let X be a (tradeable) asset whose value at $t=0$ is $X(0)$ and whose values at $t=1$ are $X(1,\uparrow)$ and $X(1,\downarrow)$, depending on whether the upstate or downstate occurs at $t=1$. From (2.12),

$$X(0) = \frac{1}{R}[\pi X(1,\uparrow) + (1-\pi)X(1,\downarrow)] \equiv \frac{1}{R}\mathbf{E}^\pi[X(1)]. \tag{2.15}$$

For the asset X we can define the **return** r_X by

$$r_X = \frac{X(1) - X(0)}{X(0)}, \tag{2.16}$$

which is shorthand for

$$r_X(\uparrow) = \frac{X(1,\uparrow) - X(0)}{X(0)}$$

$$r_X(\downarrow) = \frac{X(1,\downarrow) - X(0)}{X(0)}.$$

Lemma 2.11. *For any $0 \leq p, q \leq 1$ suppose there are associated probabilities. Then*

$$\mathbf{E}^p[r_X] - \mathbf{E}^q[r_X] = (p-q)\left[\frac{X(1,\uparrow) - X(1,\downarrow)}{X(0)}\right]. \tag{2.17}$$

Proof. Exercise. □

Remark 2.12.

$$\mathbf{E}^\pi[r_X] = r \equiv R - 1 \tag{2.18}$$

Proof. Exercise. □

Corollary 2.13.

$$\boldsymbol{E}^p\left[r_X\right] - r = (p - \pi)\left[\frac{X(1,\uparrow) - X(1,\downarrow)}{X(0)}\right] \tag{2.19}$$

$$\boldsymbol{E}^p\left[r_X\right] - r_X(\uparrow) = (p - 1)\left[\frac{X(1,\uparrow) - X(1,\downarrow)}{X(0)}\right] \tag{2.20}$$

$$\boldsymbol{E}^p\left[r_X\right] - r_X(\downarrow) = p\left[\frac{X(1,\uparrow) - X(1,\downarrow)}{X(0)}\right] \tag{2.21}$$

Proof. For (2.19), use (2.18) and $q = \pi$ in (2.17). For (2.20), use $q = 1$ in (2.17). For (2.21), use $q = 0$ in (2.17). □

Definition 2.14. *Given probability p, let X and Y be two (tradeable) assets. Their values at $t = 0$ are $X(0)$, $Y(0)$. At $t = 1$ in the $\uparrow$ state (resp., $\downarrow$ state) their values are $X(1,\uparrow), Y(1,\uparrow)$ (resp., $X(1,\downarrow), Y(1,\downarrow)$). Then define $V_{X,Y}^p$ by*

$$\begin{aligned} V_{X,Y}^p &= \boldsymbol{Cov}^p\left(r_X, r_Y\right) \\ &\equiv \boldsymbol{E}^p\left[\left(\boldsymbol{E}^p\left[r_X\right] - r_X\right)\left(\boldsymbol{E}^p\left[r_Y\right] - r_Y\right)\right] \tag{2.22} \\ &= \boldsymbol{E}^p\left[r_X r_Y\right] - \boldsymbol{E}^p\left[r_X\right]\boldsymbol{E}^p\left[r_Y\right] \tag{2.23} \end{aligned}$$

Lemma 2.15.

$$V_{X,Y}^p = p(1-p)\left[\frac{X(1,\uparrow) - X(1,\downarrow)}{X(0)}\right]\left[\frac{Y(1,\uparrow) - Y(1,\downarrow)}{Y(0)}\right] \tag{2.24}$$

Proof. Use (2.22), (2.14) together with (2.20) and (2.21). We leave the details as an exercise. □

Corollary 2.16. *The variance of X is then*

$$\sigma_X^2 \equiv V_{X,X}^p = p(1-p)\left[\frac{X(1,\uparrow) - X(1,\downarrow)}{X(0)}\right]^2. \tag{2.25}$$

Let us now assume (wlog) that $\boldsymbol{E}^p\left[r_X\right] \geq r$. With this assumption we have the following lemma.

Lemma 2.17. *Suppose that $0 < p < 1$. Then*

$$\boldsymbol{E}^p\left[r_X\right] - r = \frac{|p - \pi|}{\sqrt{p(1-p)}}\,\sigma_X \tag{2.26}$$

Proof. This follows from (2.19) and (2.25) and the assumption.

2.2 Why Is π Called a Risk Neutral Probability?

Remark 2.18. Equation (2.26) says something about the expected return from asset X in terms of its volatility (variance). We say that an asset is **riskier** when it has a higher volatility (and hence a higher value of σ_X). By (2.26), if the volatility is zero, then the expected return is just r (the risk free interest), but when the volatility is non-zero we have a higher expected return. This result fits well with reality—if you want a higher expected return you must take on more risk. However, there is one situation where this does not hold. This is when $p = \pi$. In this case your expected return is always r no matter what risk. If your (subjective) probabilities about events at $t = 1$ coincide with π, then you are insensitive to risk, or what is the same thing, you are **risk-neutral**. So π is the upstate probability of a risk neutral person.

Remark 2.19. Equation (2.15) is the usual pricing equation for an asset X, expressing $X(0)$ in terms of the future values of X via the risk neutral probability π. We can also express $X(0)$ via the subjective probability p. In fact suppose the assumption before Lemma 2.17 holds, and

$$\Lambda(p) = \frac{|p - \pi|}{\sqrt{p(1-p)}}.$$

Then by a simple rearrangement

$$X(0) = \frac{\mathbf{E}^p[X(1)]}{R + \Lambda(p)\,\sigma_X}, \qquad (2.27)$$

so the discounting must be **risk adjusted** if you use subjective probabilities. Note that $\Lambda(\pi) = 0$.

Another rearrangement starts with

$$\mathbf{E}^p[r_X] - r = \beta_{X,Y}\left[\mathbf{E}^p[r_Y] - r\right]. \qquad (2.28)$$

Here

$$\beta_{X,Y} = \frac{V^p_{X,Y}}{V^p_{Y,Y}},$$

which is a regression coefficient for the returns of X onto those of Y. This quantity is called a **beta** in financial circles, and betas are often published information. It is often the case that betas do not change too quickly from time to time. The identity (2.28) follows from (2.19) applied to both X and Y together with (2.24) and (2.25). It is necessary to consider $p \neq \pi$ and $p = \pi$ separately to avoid dividing 0 by 0, which is even invalid in finance! Equation (2.28) looks very much like the **CAPM** formula (CAPM = Capital

Asset Pricing Model), widely used in finance despite its restricted validity. It is valid in our simple model! Equation (2.28) can be arranged to give

$$X(0) = \frac{\mathbf{E}^p\left[X(1)\right]}{R + \beta_{X,Y}\left[\mathbf{E}^p\left[r_Y\right] - r\right]}, \tag{2.29}$$

which is a **relative** pricing formula using subjective probabilities. Given information about Y you can price X provided you also know the correlations between the returns on X and Y (which one sometimes assumes are relatively constant). It is because of the arrangement (2.29) that (2.28) is termed CAPM (read as CAP M). In practice Y is often related to some **index**.

2.3 More on Arbitrage

There are **two forms** of arbitrage opportunities. We suppose neither type exists in efficient markets. If they did exist they would exist only temporarily. An arbitrageur is someone who looks out for such opportunities and exploits them when they do exist.

The **type one** arbitrage opportunity arose in the proof of equation (2.6) in the last section. Indeed, if equation (2.6) did not hold we were able to make a profit at $t = 0$ without any unfunded liabilities at $t = 1$. Here one ends up with a profit at $t = 1$ in all states of the world.

The **type two** arbitrage opportunity arose in Examples 2.9 and 2.10. This is the situation where you start with nothing at $t = 0$, you have no liabilities at $t = 1$, but in one or more states of the world you can make a positive profit.

We now give some more examples:

Example 2.20 (Refer to Example 2.1). Here we exhibit a type two arbitrage. We choose S as in Example 2.1, but suppose

$$B(0) = 1 \text{ and } B(1,\uparrow) = B(1,\downarrow) = R = \frac{4}{3}.$$

Note that condition (2.1) is violated.

Choose $H_0 = 5$ and $H_1 = -1$, then $H_0 + H_1 S(0) = 0$.

At $t = 0$ we short sell one stock and invest the proceeds in a bank.

At $t = 1$ in $\uparrow$ our position is $H_0 R + H_1 S(1,\uparrow) = 0$; in other words our investment gives rise to $5 \times \frac{4}{3} = \frac{20}{3}$, which is enough to cover the repurchase of stock at $\frac{20}{3}$, which is then returned to its owner.

At $t = 1$ in $\downarrow$ our position is $H_0 R + H_1 S(1,\uparrow) = \frac{20}{9}$; in other words our investment gives rise to $5 \times \frac{4}{3} = \frac{20}{3}$, which is enough to cover the repurchase of stock at $\frac{40}{9}$, which is then returned to its owner, and with $\frac{20}{9}$ to spare.

Remark 2.21. Type two non-arbitrage was also used in Example 2.10.

Exercise 2.22 (A Variant of the One Price Theorem). Let X and Y be two assets (or portfolios of assets). Prove

1. if $X(1) > Y(1)$, then $X(0) > Y(0)$;
2. if $X(1) = Y(1)$, then $X(0) = Y(0)$.

Remark 2.23. $X(1) > Y(1)$ has the meaning of the notation on page 20. That is: $X(1) \geq Y(1)$ in all states of the world at $t = 1$, and strict inequality holds in at least one state of the world. [It is possible we are thinking beyond binomial models here.]

In fact, if 1. does not hold, we can obtain a type two arbitrage by short selling X and buying Y; if 2. does not hold, we obtain a type one arbitrage. The principle is this: **(short) sell high, buy low**.

2.4 The Model of Cox-Ross-Rubinstein

We shall now describe the Cox-Ross-Rubinstein model and we shall write **CRR** for **Cox-Ross-Rubinstein**. See [18].

The following notation will be used:

$$S(0) = S > 0$$
$$S(1,\uparrow) = uS$$
$$S(1,\downarrow) = dS,$$

where, as in equation (2.1),

$$0 < d < R < u.$$

Then

$$\pi = \frac{RS(0) - S(1,\downarrow)}{S(1,\uparrow) - S(1,\downarrow)} = \frac{R - d}{u - d} \qquad (2.30)$$

$$1 - \pi = \frac{S(1,\uparrow) - RS(0)}{S(1,\uparrow) - S(1,\downarrow)} = \frac{u - R}{u - d} \qquad (2.31)$$

and

$$X(0) = \frac{1}{R}\left[\frac{R-d}{u-d}X(1,\uparrow) + \frac{u-R}{u-d}X(1,\downarrow)\right]. \qquad (2.32)$$

26   2 The Binomial Model for Stock Options

Example 2.24 (European call option). Here $X(1) = (S(1) - K)^+$.
Assume $S(1,\downarrow) < K < S(1,\uparrow)$, then

$$X(1,\uparrow) = (S(1,\uparrow) - K)^+ = uS - K$$
$$X(1,\downarrow) = (S(1,\downarrow) - K)^+ = 0$$

and so

$$X(0) = \frac{\pi(uS - K)}{R}$$
$$= S\left[\frac{\pi u}{R}\right] - \left[\frac{K}{R}\right]\pi. \qquad (2.33)$$

Remark 2.25. For those familiar with the Black and Scholes formula for pricing call options,

$$C(0) = S(0)\mathcal{N}(d_1) - Ke^{-rT}\mathcal{N}(d_2), \qquad (2.34)$$

we note an obvious similarity. Here $\mathcal{N}$ has the definition

$$\mathcal{N}(x) \equiv \frac{1}{\sqrt{2\pi}}\int_{-\infty}^{x} e^{-\frac{1}{2}y^2}\, dy. \qquad (2.35)$$

We shall meet these ideas again later. The expressions for d_1 and d_2 are given in Chapter 4.

Continuing, we note that

$$0 < \frac{\pi u}{R} < 1. \qquad (2.36)$$

In fact

$$0 < \frac{\pi u}{R} = \left(\frac{R-d}{u-d}\right)\frac{u}{R} = \frac{1 - \frac{d}{R}}{1 - \frac{d}{u}} < 1$$

as $R < u$ implies $1 - \frac{d}{R} < 1 - \frac{d}{u}$.
If $K \leq S(1,\downarrow)$, then $X(1) = S(1) - K$ and so

$$X(0) = \frac{\pi(S(1,\uparrow) - K) + (1-\pi)(S(1,\downarrow) - K)}{R}$$
$$= \frac{\pi S(1,\uparrow) - +(1-\pi)S(1,\downarrow)}{R} - \frac{K}{R}$$
$$= S(0) - \frac{K}{R}.$$

If $K \geq S(1,\uparrow)$, then $X(1) = 0$ and so $X(0) = 0$.

Example 2.26. Consider the claim $X(1) = (K - S(1))^+$. This is a European put option in the binomial model. Assume $S(1,\downarrow) < K < S(1,\uparrow)$; then

$$X(1,\uparrow) = (K - S(1,\uparrow))^+ = 0$$
$$X(1,\downarrow) = (K - S(1,\downarrow))^+ = K - dS$$

and so

$$X(0) = \frac{(1-\pi)(K - dS)}{R}$$
$$= \left[\frac{K}{R}\right](1-\pi) - S\left[\frac{(1-\pi)d}{R}\right]. \tag{2.37}$$

Remark 2.27. As mentioned before, π is called a **risk-neutral probability** (of being in state $\uparrow$). It is characterized by

$$S(0) = \frac{\pi S(1,\uparrow) + (1-\pi)S(1,\downarrow)}{R}.$$

This says that under π, the expected discounted value of $S(1)$ is $S(0)$.

2.5 Call-Put Parity Formula

This is also called **put-call parity**. It applies to **European** style call and put options.

There are several **model-independent** formulae in finance. Clearly, such formulae are very important. We shall meet a number of them. The most well known one is the call-put parity formula, which states:

$$C(0) - P(0) = S(0) - \frac{K}{R}, \tag{2.38}$$

at least in the present framework. We shall discuss generalizations later.

The calls and puts in this formula are assumed to have the same strike price K and the same time to expiry (maturity).

CRR Model-Dependent Proof

Suppose that $S(1,\downarrow) < K < S(1,\uparrow)$. Then with $S = S(0)$ and (2.33) and (2.37),

$$C(0) = \frac{\pi}{R}[uS - K]$$

$$P(0) = \frac{1-\pi}{R}[K - dS]$$

$$C(0) - P(0) = \frac{\pi}{R}[uS - K] - \frac{1-\pi}{R}[K - dS]$$

$$= \frac{\pi(uS) + (1-\pi)(dS)}{R} - \frac{\pi K + (1-\pi)K}{R}$$

$$= S(0) - \frac{K}{R}$$

By the way, $\frac{K}{R} = PV(K) \equiv PV_0(K)$, the present value at $t = 0$ of K at $t = 1$ (PV = Present Value).

Model-Independent Proof. Model-independent relations are very important.

We again have two times: now ($t = 0$), and expiry date ($t = T$). Assume (if possible) that

$$C(0) - P(0) - S(0) + PV(K) > 0. \tag{2.39}$$

We shall show there is type one arbitrage. At $t = 0$, we short sell a call option, buy a put option, buy one stock, borrow $PV(K)$. The short sale and the borrowing is enough to cover the put options and stock price, and there is cash left over (by (2.39)), which we pocket.

At expiry ($t = T$), we **cash settle the call**, realize value of the put, sell the stock, repay the loan. The net of all these transactions is

$$-(S(T) - K)^+ + (K - S(T))^+ + S(T) - K = 0. \tag{2.40}$$

The person who let you borrow the call only needs the cash value $((S(T) - K)^+)$ of the call at expiry (called **cash settling**). The assets (put and stock), are just enough to cover the liabilities of the call and loan repayment.

One can demonstrate (2.40) by looking at the two cases: $S(T) > K$ and $S(T) \leq K$. In the first case:

$$-(S(T)-K)^+ +(K-S(T))^+ +S(T)-K = -(S(T)-K)+0+S(T)-K = 0$$

and in the latter

$$-(S(T)-K)^+ +(K-S(T))^+ +S(T)-K = 0+(K-S(T))+S(T)-K = 0.$$

The other case of (2.39),

$$C(0) - P(0) - S(0) + PV(K) < 0,$$

is treated in a similar way. First write this as

$$-C(0) + P(0) + S(0) - PV(K) > 0. \qquad (2.41)$$

At $t = 0$, buy a call option, short sell a put option, short sell a stock and invest $PV(K)$. The two short sales are enough to cover the call options and the investment amount. Further, there is cash left over (by (2.41)), which we pocket.

At expiry ($t = T$), we realize value of the call, cash settle the put, buy a stock and return, realize the investment K. The net cost of all these transactions is

$$(S(T)-K)^+ - (K-S(T))^+ - S(T) + K = 0 \qquad (2.42)$$

as before.

In both cases, we can pocket a profit at $t = 0$ and have no unfunded liabilities at expiry. These are type one arbitrages. These financial contradictions show the call-put parity equality must hold. [A reason to prefer the term call-put parity is because it could also be read "call minus put" which is the left hand side of the call-put parity formula. It reminds us which way they are around!].

2.6 Non Arbitrage Inequalities

In the section above we saw the first of these: the call-put parity formula. This was proved in the CRR one-step model and then we gave a model independent proof. It is the fact that it has a model independent proof which makes it a fundamental result. However, note that the call-put parity formula holds for European options. It does not hold for the American style counterparts.

We now investigate other results for which there are model-independent proofs. Consequently, we are no more in the simple two-state, one-period model.

Example 2.28 (Lower bounds for European calls).

Let $0 \leq t < T$. Let $C(t)$ be the value at t of a European call option that expires at time T, whose strike price is K. We also write $PV_t(K)$ for the value at time t of K at time T. This amount could be found by some discounting formula whose precise details do not matter here—as long as interest rates are not random. Then

$$C(t) \geq [S(t) - PV_t(K)]^+ \qquad (2.43)$$
$$= \max[0, S(t) - PV_t(K)]. \qquad (2.44)$$

Proof. Clearly $C(t) \geq 0$ as $C(T) = (S(T) - K)^+ \geq 0$. [See Exercise (2.22)] So we only need to show

$$C(t) \geq S(t) - PV_t(K). \qquad (2.45)$$

Suppose to the contrary that

$$C(t) < S(t) - PV_t(K),$$

which is the same as

$$S(t) - PV_t(K) - C(t) > 0. \qquad (2.46)$$

If (2.46) were the case, we show how to create an arbitrage.

At time t we short sell one stock, invest $PV_t(K)$ in a bank, buy a call option, (expiring at T with strike price K). The short sale is enough to cover the purchases and (2.46) says there is a positive amount left over for the pocket.

At time T, the expiry date of the call option, we buy a stock and return it, realize the value of the call, realize the value of the investment in the bank, (take the K out of bank). The net proceeds are given by the left hand side of

$$-S(T) + (S(T) - K)^+ + K \geq 0. \qquad (2.47)$$

This implies that there are no unfunded liabilities at time T. To show (2.47) we consider two cases: $S(T) > K$ and $S(T) \leq K$. For the former

$$-S(T) + (S(T) - K)^+ + K = -S(T) + (S(T) - K) + K = 0$$

and in the latter case

$$-S(T) + (S(T) - K)^+ + K = -S(T) + 0 + K \geq 0.$$

Therefore arbitrage has been established. This is a financial contradiction and so (2.45) and hence (2.43) and (2.44) hold. $\square$

Example 2.29 (American call options). It is not optimal to exercise an American call option before expiry if the underlying stock does not pay dividends during the life of the option. In fact, at any time t prior to expiry,

$$C_A(t) > (S(t) - K)^+. \tag{2.48}$$

In other words, before expiry, an American call option is always worth (strictly) more than its exercise value. For this we require positive interest rates.

Proof. First note that

$$C_A(t) \geq C_E(t). \tag{2.49}$$

That is, an American call option is always worth at least the same as the European counterpart. After all the American call option offers all the privileges of the European call option and other benefits besides—the right to exercise the call before expiry, for example. One can also argue more rigorously: Assume that (2.49) is not true for some time t prior to expiry and construct an arbitrage opportunity. Suppose at time t it is true that $C_E(t) - C_A(t) > 0$. At time t short sell the European call and purchase an American call (with the same specifications of strike price and exercise price). Pocket the profit. As you own the American call option you decide when to exercise it. Decide **not** to exercise it early. At expiry the realized value of the American Call Option is $(S(T) - K)^+$, which is just the same as the value of the European call option. So this realized value can be used to cash settle the European call option at time T.

This argument is included to show how non arbitrage arguments can be used to derive financial conclusions.

Suppose now that $S(t) > K$. Then, as interest rates are positive,

$$S(t) - PV_t(K) > S(t) - K > 0. \tag{2.50}$$

So (using Example 2.28)

$$\begin{aligned} C_A(t) &\geq C_E(t) \\ &\geq S(t) - PV_t(K) \\ &> S(t) - K = (S(t) - K)^+. \end{aligned}$$

Thus, (2.48) holds if $S(t) > K$.

Suppose now that $S(t) \leq K$; then $(S(t) - K)^+ = 0$. But $C_A(t) > 0$ for $t < T$, so again (2.48) holds if $S(t) \leq K$. □

Remark 2.30. 1. One consequence of this example is the following: If a stock does not pay a dividend during the life of an option, then the American call options and the European call option have the same value. As dividends are usually paid twice a year, there will be many short term call options (90-day options) for which this condition applies. The financial press usually advertises when dividends are paid, and sometimes predicts when the next dividends will be paid based on what happened the year before.

2. The corresponding result does not hold for American and European put options. Always before expiry at $t < T$

$$P_A(t) > P_E(t), \qquad (2.51)$$

where we have assumed that the two puts are the same in every other respect. The difference

$$e(t) \equiv P_A(t) - P_E(t) > 0 \qquad (2.52)$$

is called the **early-exercise premium**. This is the extra amount one pays for an American put to have the right to exercise it early.

Example 2.31 (Estimate interest from option prices). For European calls and puts, the call-put parity formula can be rearranged to yield

$$R = \frac{K}{S(0) + P(0) - C(0)}. \qquad (2.53)$$

For American puts and calls we have

$$C_A(0) - P_A(0) \le S(0) - PV(K) \qquad (2.54)$$

when dividends are not paid during the life of the options. This can be deduced from the call-put parity formula for European options and using $C_A(0) = C_E(0)$ and $P_A(0) \ge P_E(0)$.

Proof. Exercise. □

Under these circumstances (regarding dividends) we have

$$R \ge \frac{K}{S(0) + P_A(0) - C_A(0)}. \qquad (2.55)$$

We should be able to check this holds from data in the financial press (otherwise there are arbitrage opportunities)

2.6 Non Arbitrage Inequalities 33

Example 2.32 (An AOL example). Time $t = 0$ is 22 July 2003 (the previous trading day). $S(0) = \$16.85$, $K = \$16.00$, $C(0) = \$1.20$, $P(0) = \$0.45$. Then the right hand side of (2.55) is

$$\frac{16.00}{16.85 + 0.45 - 1.20} = 0.9937. \tag{2.56}$$

The interest rate at the time was around 1.00%, so there is no violation of (2.55).

Remark 2.33. We have seen that an American call and a European call have the same value when there are no dividends paid on the underlying stock. Under these circumstances we saw that it is not optimal to exercise an American call option early, as it is more profitable to sell the option than to exercise it. So the early exercise feature under these circumstances provides no extra value. We shall discuss later what happens when there are dividend payments.

Example 2.34 (Call options are decreasing functions of their time to expiration). Suppose there are two calls which are identical except they have different expiration times T_1 and T_2, with $T_1 > T_2$. "Now" is time t. Their times to expiration are τ_1 and τ_2 with $\tau_i = T_i - t, i = 1, 2$, so $\tau_1 > \tau_2$. The values of these calls at time t are $C^{\tau_i}(t)$, for $i = 1, 2$. We claim

$$C^{\tau_1}(t) \geq C^{\tau_2}(t) \tag{2.57}$$

for $0 \leq t \leq T_2$. (After T_2 the call with shorter time to expiry ceases to exist so (2.57) is either obvious or does not mean much, depending how you view things.) To prove (2.57) assume (if possible) that

$$C^{\tau_2}(t) - C^{\tau_1}(t) > 0$$

for some $0 \leq t \leq T_2$. This leads to an arbitrage opportunity as follows. At t, short the τ_2 call and purchase the τ_1 call. Pocket the profit. At time T_2, the value $V(T_2)$ of the position is

$$\begin{aligned} V(T_2) &= -(S(T_2) - K)^+ + C^{\tau_1}(T_2) \\ &\geq -(S(T_2) - K)^+ + (S(T_2) - K)^+ \\ &= 0, \end{aligned}$$

where we have used Example 2.28. So at time T_2 we have no unfunded liabilities if we sell the longer-dated call and cash settle the shorter-dated one at this time. So we now have a type one arbitrage opportunity. This is a (financial) contradiction, and so our claim holds.

Many more relations can be deduced in this model-independent way.

2.7 Exercises

Exercise 2.35. This exercise refers to Example 2.24. What are the reasons that market players will buy and sell put options. Are buyers and sellers matched?

Exercise 2.36. In Example 2.2 show that $0 \leq H_1 \leq 1$.

Exercise 2.37. Prove the identities (2.17), (2.18), (2.26), (2.27) and (2.28) in Section 2.2.

Exercise 2.38. This exercise refers to Example 3.1. How can you decide whether a futures trader is a speculator or a hedger? Explain the market for futures/forwards.

Exercise 2.39. Let $S = \{S(t) \mid t \geq 0\}$ be the price process of some stock (e.g., AOL shares). Let $C(t)$ denote the value at time t of a (European) call option written on S with maturity date T and exercise price K. Then $C(T) = \max[0, S(T) - K]$. Draw a graph, plotting $C(T)$ versus $S(T)$. This is called the payoff graph for the this call option. The profit graph is the plot of $C(T) - C(0)$ versus $S(T)$. Draw the profit graph. For what values of $S(T)$ will the profit be positive? (This profit ignores the time value of money.)

Exercise 2.40. Repeat Exercise 2.39 but for the (European) put option. The difference between a put and a call is that with the put you have the right to sell rather than the right to buy. If $P(t)$ is the value of this put with strike price K and expiry date T, explain why $P(T) = \max[0, K - S(T)]$. Plot $P(T)$ versus $S(T)$, and $P(T) - P(0)$ versus $S(T)$. If you want to take a numerical example, choose the AOL/AUG03/16.00/PUT with $P(0) = 0.45$USD. For what values of $S(T)$ will the profit be positive? Explain why the holding of a put option on S is like holding an insurance policy over S.

Exercise 2.41. The current price of a certain stock is $94 and 3-month call options with a strike price of $95 currently sell for $4.70. An investor who feels that the stock price will increase is trying to decide between buying 100 shares and buying 2000 call options (20 contracts). Both strategies would involve an investment of $9,400. What advice would you give the investor? How high does the stock price have to rise for the option strategy to be more profitable?

Exercise 2.42. Suppose two banks XYZ and ABC are equally rated (as regards risk). Suppose that XYZ offers and charges customers 4% interest on deposits or loans, while ABC offers and charges 6% interest. Seeing this situation, how could you make a riskless profit without using any of your own money? You should provide an explicit strategy for achieving this, and explain any problems you might have carrying it out in practice. If the two banks were not equally rated, what possible reason could you give for the difference in interest rates?

Exercise 2.43. Suppose IBM pays a dividend D on their shares S at time τ. Show that $S(\tau+) = S(\tau-) - D$. Actually, to be precise, τ should be what is called the ex-dividend date. You should again argue your solution from the assumption of no arbitrage. $S(\tau+)$ means the value of S just after τ, and $S(\tau-)$ the value just before.

Exercise 2.44. Let C_i for $i = 1, 2, 3$ be European call options all expiring at T with strike prices K_i, for $i = 1, 2, 3$ all written on the same stock S. The **butterfly spread** is the combination $C_1 - 2C_2 + C_3$ with $K_2 = \frac{1}{2}(K_1 + K_3)$. Graph $C(T)$ against $S(T)$. Show that $C_2(0) < \frac{1}{2}(C_1(0) + C_3(0))$. Discuss which assumptions you make.

Exercise 2.45. We established call-put parity formula holds for European call and put options:

$$C(0) - P(0) = S(0) - PV(K),$$

where $PV(K) = K/R$. With the choices $S(0) = \$10.50$, $K = \$10.00$, $C(0) = \$3.00$, $P(0) = \$1.00$ and $R = 1.0043$, show that the call-put parity formula is violated. Show how to create an arbitrage opportunity of at least $\$1000$. You must not use any of your own money to fund this arbitrage opportunity.

Exercise 2.46. Read Linear Regression in the Appendix to Exercise 2.47, which can be studied together with this exercise.

Consider the data in Table 2.1 for XYZ/AUG03/CALLs. Suppose the spot price of the XYZ shares is $S = \$16.96$. (Not all strike prices are used here.)

Table 2.1. XYZ/AUG03/CALL for Exercise 2.46

	1	2	3	4	5	Sum
n_i	171	316	475	802	594	$n =$
$w_i = \frac{n_i}{n}$						1
x_i	17.50	18.00	18.50	19.00	19.50	
y_i	0.36	0.19	0.12	0.06	0.06	

We had $N = 5$, the x values represent strike prices, and the y values represent (ask) call prices. The n values are for open interest, which gives the number of contracts presently held with a particular strike price. We plot call prices against strike prices and seek the least squares fit line. Find its slope m and intercept c. Using the equations

$$c = \frac{\pi u S}{R} \qquad m = -\frac{\pi}{R} \qquad d = \frac{R - \pi u}{1 - \pi},$$

estimate π, u and d.

Use these values to compute the value of the XYZ/AUG02/17.00/PUT and compare your answer with the market value 0.55 CAD (ask).

These XYZ options are really American options, but as they are rarely exercised (why?) they price like European options. The value of R is to be taken as 1.0042.

Exercise 2.47. Consider the one-step binomial model with stock prices having $d = 1/u$. We can price ATM calls by

$$c = \frac{\pi}{R}(uS - K) = \frac{\pi}{R}(u-1)S.$$

This leads to

$$u = \frac{1 + Rx}{R(1-x)},$$

where $x = c/S$. So u can be calculated for a range of ATM calls by this formula. The CRR paper uses $u = \exp(\sigma\sqrt{\Delta t})$ where σ is the volatility and Δt is the time interval between $t = 0$ and $t = 1$, which we will take as $31/365 = 0.08493$ and so $\sqrt{\Delta t} \approx 0.291431$. It may therefore be of interest to plot $\ln u$ versus σ, get the line of best fit, and see if the estimated slope is about 0.29143. This is what you are now asked to do. Here are some "data" for various AUG03 calls. Use $R = 1.0042$ as before.

Table 2.2. ATM Call Prices for Exercise 2.47

Company	S	ATM Call Price	σ
ABC	16.96	0.6800	0.2685
DEF	9.18	0.3567	0.3614
HIJ	29.10	0.9780	0.2345
KLM	31.00	1.0400	0.2841
NOP	8.46	0.6800	0.6765

In making your line of least squares fit, use $w_i = 1/5 = 0.2$ for $i = 1, 2, 3, 4, 5$. Also observe that you should use the zero intercept form of linear regression here.

Appendix to Exercises 2.46 and 2.47

Linear Regression

We are given data points $\{(x_i, y_i) \mid i = 1, 2, \ldots, N\}$ and we want to place a line of best fit through them. The model will be

$$y_i = mx_i + c + \epsilon_i \tag{2.58}$$

for each $i = 1, 2, \ldots, N$. Here ϵ_i denotes an error for each i.
The **least squares fit line** is that line (or choice of m, c) so that

$$\sum_{i=1}^{N} \epsilon_i^2 \qquad (2.59)$$

is minimal.

In finance we may have n_i measurements all at (x_i, y_i) and n_i may not be the same for each i. For example we could plot call prices versus strike prices from NYSE data. For the n_i we could use the open interest or the volume of trade.

In either case let us put

$$M = n_1 + n_2 + \ldots + n_N$$

and set

$$w_i \equiv \frac{n_i}{M},$$

which gives the proportion of measurements at (x_i, y_i). We could then minimize

$$\sum_{i=1}^{N} w_i \epsilon_i^2. \qquad (2.60)$$

Setting derivatives of this expression with respect to m and c, both to zero yields the estimates for m and c which are

$$m = \frac{\overline{xy} - \overline{x}\,\overline{y}}{\overline{x^2} - \overline{x}^2} \qquad (2.61)$$

and

$$c = \overline{y} - m\overline{x}. \qquad (2.62)$$

Here we are using

$$\overline{y} = \sum_{i=1}^{N} w_i y_i$$

38   2 The Binomial Model for Stock Options

$$\overline{x} = \sum_{i=1}^{N} w_i x_i$$

$$\overline{xy} = \sum_{i=1}^{N} w_i x_i y_i$$

$$\overline{x^2} = \sum_{i=1}^{N} w_i x_i^2.$$

If $N > 1$, then $\overline{x^2} \neq \overline{x}^2$, so we never divide by zero! In fact

$$\overline{x^2} - \overline{x}^2 = \sum_{i=1}^{N} w_i (x_i - \overline{x})^2$$

$$\overline{xy} - \overline{x}\,\overline{y} = \sum_{i=1}^{N} w_i (x_i - \overline{x})(y_i - \overline{y}),$$

so we can solve for m and hence for c.

Example 2.48 (linear regression). Consider some XYZ call option prices. Suppose N = 5; $x_1 = 10.50$, $x_2 = 11.00$, $x_3 = 11.50$, $x_4 = 12.00$, $x_5 = 12.50$ and $y_1 = 1.36$, $y_2 = 0.95$, $y_3 = 0.62$, $y_4 = 0.38$, $y_5 = 0.20$ (we are using the selling prices). We could weight by **open interest** (open interest is the number of contracts in a particular class of options), and suppose $n_1 = 56$, $n_2 = 662$, $n_3 = 941$, $n_4 = 969$, $n_5 = 268$. Then $M = 2896$ and so $w_1 = \frac{56}{2896}$, $w_2 = \frac{662}{2896}$, $w_3 = \frac{941}{2896}$, $w_4 = \frac{969}{2896}$, $w_5 = \frac{268}{2896}$. Then

$$\overline{x} = \frac{56 \times 10.50 + 662 \times 11.00 + 941 \times 11.50 + 969 \times 12.00 + 268 \times 12.50}{2896}$$
$$= 11.62620856$$

$$\overline{y} = \frac{56 \times 1.36 + 662 \times 0.95 + 941 \times 0.62 + 969 \times 0.38 + 268 \times 0.20}{2896}$$
$$= 0.590573204$$

$$\overline{x^2} = \frac{56 \times 10.50^2 + 662 \times 11.00^2 + \ldots + 969 \times 12.00^2 + 268 \times 12.50^2}{2896}$$
$$= 135.4054731$$

$$\overline{xy} = \frac{56 \times 1.36 \times 10.50 + 662 \times 0.95 \times 11.00 + \ldots + 268 \times 0.20 \times 12.50}{2896}$$
$$= 6.73879489$$

$$m = \frac{\overline{xy} - \overline{x}\,\overline{y}}{\overline{x^2} - \overline{x}^2}$$

$$= \frac{6.73879489 - 11.62620856 \times 0.590573204}{135.4054731 - 11.62620856^2}$$

$$= \frac{-0.127772349}{0.23674762} = -0.539698557$$

$$c = \overline{y} - m\overline{x} = 0.590573204 + 0.539698557 \times 11.62620856$$
$$= 6.865221193.$$

We shall see later some interpretation of these estimates of m and c. If you know some statistics about the errors, then you can discuss the confidence intervals of the estimators of m and c given in formulae (2.61) and (2.62).

Many of these calculations can be easily carried out in MS-EXCEL.

Zero-Intercept Linear Regression

The model will now be

$$y_i = mx_i + \epsilon_i \qquad (2.63)$$

for each $i = 1, 2, \ldots, N$. Here ϵ_i denotes an error for each i.

The (weighted) **least squares fit line** is that line (or choice of m) making

$$\sum_{i=1}^{N} w_i \epsilon_i^2 \qquad (2.64)$$

minimal. Setting the derivative of this expression with respect to m to zero gives:

$$m = \frac{\overline{xy}}{\overline{x^2}} \qquad (2.65)$$

with the same notation as above.

Exercise 2.49. Show that the value of a call can never be less than the value of an otherwise identical call with a higher strike price; that is,

$$C(K_1) \geq C(K_2) \quad \text{if } K_2 > K_1,$$

and furthermore

$$K_2 - K_1 \geq C(K_1) - C(K_2) \quad \text{if } K_2 > K_1.$$

3

The Binomial Model for Other Contracts

We now discuss the simple **one-step binomial model** applied to other financial contracts.

3.1 Forward Contracts

We discuss two related contracts: the forward contract and the futures contract. The latter is traded on a futures exchange, (e.g. the CBOE), the former between two parties. We shall discuss the futures contract later.

Definition 3.1 (Forward Contract). *A (long/short) forward contract is an agreement to buy or sell an asset (S, say) at some future time (T, say) for an agreed price (F, say). There are no payments made initially (at time $t = 0$), and the price F is called the **delivery price**.*

Therefore, a **long forward contract** is a binding agreement to buy, while a **short forward contract** is a binding agreement to sell. The word "**binding**" distinguishes this contract from the option contracts.

The **payoff** of a **long forward contract** is then $S(T) - F$. (Buy at F and sell at $S(T)$). This payoff may be positive, negative or zero depending on the state of the world at time T. No payments are made initially, at $t = 0$, as the present value of the forward contract is zero. (We ignore transaction costs.) The agreed price F is called the **forward price**. The rational value of F can be shown, in a model-dependent way or in a model-independent way, to be $F = S(0)R$. In the previous chapter we talked about $t = 1$ without specifying units of time. In this context T denotes the time interval between $t = 0$ and $t = 1$, and R is the value at T of 1 at the present time.

Proof (Model-Dependent). Consider the one-step binomial asset pricing model. Suppose at time T, $S(T)$ equals either $S(T,\uparrow)$ or $S(T,\downarrow)$. The time $t = 0$ value of the forward contract is zero, so

$$\begin{aligned} 0 &= \frac{1}{R}\left[\pi\left(S(T,\uparrow) - F\right) + (1-\pi)\left(S(T,\downarrow) - F\right)\right] \\ &= \frac{1}{R}\left[\pi S(T,\uparrow) + (1-\pi)S(T,\downarrow)\right] - \frac{F}{R} \\ &= S(0) - \frac{F}{R}. \end{aligned}$$

That is

$$F = S(0)R \ .$$

□

Proof (Model-Independent). Assume (if possible) that

$$F - S(0)R > 0.$$

At time $t = 0$, borrow $S(0)$ in cash, buy one stock, enter a (short) forward contract to sell the stock for F at time T. There is a net cost of \$0 at time $t = 0$.

At expiry T, sell the stock for F and repay the loan with interest: $S(0)R$. The net position is $F - S(0)R > 0$, which is a clear profit. So, with no net outlay at time $t = 0$, one can generate a positive profit at time T. This is an arbitrage, which violates our basic axiom.

Assume now (if possible) that

$$S(0)R - F > 0.$$

A similar argument works.

At time $t = 0$, short sell one stock, invest the amount raised, $S(0)$, in a bank, enter a forward contract to buy a stock at time T for F. There is a net cost of \$0 at time $t = 0$.

At expiry T, buy a stock for F and return it. The stock purchase is funded from the investment $S(0)R$. The net position is $S(0)R - F > 0$, which is a clear profit. Again with no net outlay at time $t = 0$, one can generate a positive profit at time T. This is an arbitrage and a violation of the axiom.□

Remark 3.2. We should discuss the market for forward contracts. Why do players enter these contracts? There are again different reasons depending on whether a player is a speculator, a hedger or an arbitrageur.

Forward and futures are particularly important for **commodities**.

For a farmer, only September wheat prices may be important, for that is the time that the crop is ready for harvest and sale. It is at this time that a food company must buy (durum) wheat for pasta manufacture. Farmers and a company could enter into forward contracts. The farmers could agree to sell a given number of bushels of a particular grade of wheat to the company for an agreed price. From the farmer's point of view the income from the wheat sales will be known ahead of the harvest. From the company's point of view the cost of wheat input into the pasta production will be known ahead of time. The forward contract has taken the uncertainty out of the income for the farmer and the uncertainty out of the input costs for the company.

At expiry T of the forward contract either party could be worse off compared with not having a forward contract. Suppose that $S(T) > F$. Then the farmer must sell the wheat for F and not for the higher price $S(T)$, and so suffer a relative loss. In this case the company will have a relative gain (paying F rather than $S(T)$). However this is not the point. By entering a forward contract, one **gives up** the possibility of gain to obtain a certain outcome.

This situation is not without risks, however. If there is a poor harvest (a drought, for example), then the farmer may not be able to deliver the agreed amount. In that case, the farmer must go to the market to buy the shortfall (at price $S(T)$) and deliver it to the company for F. Because of this, the farmer would probably only forward sell 70% (say) of the wheat crop this way. Another risk could be that the **grade** of the wheat harvested is not up to the **quality** agreed in the forward contract. Again the farmer may suffer a loss in order to honor the forward contract.

A more complete discussion of forwards and futures is given in Duffie [26] and in many other books.

3.2 Contingent Premium Options

The put and call options we have described so far are called (plain) **vanilla options**. The opposite of vanilla is **exotic**. However, some derivatives which were once called exotic are now regarded as vanilla. This applies to the **barrier options** and the Asian or average rate options.

The contingent premium option is a **pay-later option**. Let us describe the European call contingent premium option. It is the same as the vanilla European call except that the premium is not paid up front (at $t = 0$) but at expiry, and then only if the option expires **in the money**.

We shall price this option in the one-step binomial asset pricing model. Write X for the price of the European call contingent premium Option. Let α be the contingent premium. Then $X(0) = 0$, as nothing is paid at time $t = 0$. Assume $S(1,\downarrow) < K < S(1,\uparrow)$.

At expiry ($t = 1$), the payoff is

$$X(1,\uparrow) = S(1,\uparrow) - K - \alpha$$
$$X(1,\downarrow) = 0.$$

Then

$$0 = X(0) = \frac{1}{R}[\pi X(1,\uparrow) + (1-\pi)X(1,\downarrow)]$$
$$= \frac{1}{R}\pi[S(1,\uparrow) - K] - \frac{\pi}{R}\alpha$$

so

$$\alpha = \frac{R}{\pi}C(0),$$

where

$$C(0) = \frac{\pi}{R}[S(1,\uparrow) - K]$$

is the price of the vanilla call. Now, as $R \geq 1$ and $0 < \pi < 1$, we see $\alpha > C(0)$, as you might have expected.

Another variant could be the following. Pay 50% of the premium at $t = 0$ and 50% at expiry ($t = 1$) provided that one is in the money. Then the premium α is determined from

$$\frac{\alpha}{2} = C(0) - \frac{\pi}{R} \cdot \frac{\alpha}{2}$$

or

$$\alpha = 2 \cdot C(0) \cdot \left[1 + \frac{\pi}{R}\right]^{-1} > C(0).$$

3.3 Exchange Rates

Let X denote an exchange rate. We shall now discuss the USD/CAD exchange rates, though any other example will do.

Let $X(t)$ denote the value at time t of 1 USD expressed in CAD (or the domestic currency). This is called the **direct** (or American), way of quoting an exchange rate and is the one that will be used in this book. It is certainly the natural way to quote exchange rates. We ask for the value in CAD of one US dollar just as we ask for the value in CAD (**domestic currency**) of a kilo of tomatoes.

For historical reasons (see **Bretton-Woods Agreement**) the market quotes some rates in an **inverse** (European) way. So we have that 1 CAD = 0.72 USD (inverse, market convention) is equivalent to 1 USD = $\frac{1}{0.72}$ = 1.389 CAD (the direct value), which we call $X(0)$, say.

We now describe the pricing formula in a one-step binomial world.

Suppose that at time $t = 1$, X can take one of two values, $X(1,\uparrow) > X(1,\downarrow)$. There will also be two interest rates, domestic (d = domestic = Canadian),

$$B_d(0) = 1, \quad B_d(1,\uparrow) = B_d(1,\downarrow) = R_d = 1 + r_d,$$

and foreign, (f = foreign = USA)

$$B_f(0) = 1, \quad B_f(1,\uparrow) = B_f(1,\downarrow) = R_f = 1 + r_f.$$

We wish to price a foreign exchange (FX) contingent claim W. In other words, given $W(1)$ in CAD, find $W(0)$ in CAD.

We again will use replication and non arbitrage concepts.

We take H_0 CAD and H_1 USD. So this portfolio has value (in CAD),

$$H_0 + H_1 X(0) \tag{3.1}$$

at time $t = 0$. At time $t = 1$, this portfolio is worth (in CAD),

$$H_0 R_d + H_1 R_f X(1).$$

We choose H_0 and H_1 so that

$$W(1) = H_0 R_d + H_1 R_f X(1). \tag{3.2}$$

This is the same as the two equations:

$$W(1,\uparrow) = H_0 R_d + H_1 R_f X(1,\uparrow) \qquad (3.3a)$$
$$W(1,\downarrow) = H_0 R_d + H_1 R_f X(1,\downarrow) \qquad (3.3b)$$

and if $X(1,\uparrow) \neq X(1,\downarrow)$, we have a unique solution for H_0 and H_1:

$$H_1 = \frac{1}{R_f} \frac{W(1,\uparrow) - W(1,\downarrow)}{X(1,\uparrow) - X(1,\downarrow)}$$

$$H_0 = \frac{1}{R_d} \frac{X(1,\uparrow)W(1,\downarrow) - X(1,\downarrow)W(1,\uparrow)}{X(1,\uparrow) - X(1,\downarrow)}.$$

This leads to

$$W(0) = \frac{1}{R_d} [\pi W(1,\uparrow) + (1-\pi) W(1,\downarrow)] \qquad (3.4)$$

with

$$\pi = \frac{\frac{R_d}{R_f} X(0) - X(1,\downarrow)}{X(1,\uparrow) - X(1,\downarrow)} \qquad (3.5)$$

$$1 - \pi = \frac{X(1,\uparrow) - \frac{R_d}{R_f} X(0)}{X(1,\uparrow) - X(1,\downarrow)}. \qquad (3.6)$$

Consequently for, $0 < \pi < 1$, our exchange rate model must satisfy

$$X(1,\downarrow) < \frac{R_d}{R_f} X(0) < X(1,\uparrow) \qquad (3.7)$$

as we assumed that $X(1,\downarrow) < X(1,\uparrow)$.

Definition 3.3 (Arrow-Debreu Securities). *These are contingent claims which equal 1 in a single future state and are 0 otherwise.*

Let $W^\uparrow$ and $W^\downarrow$ be the (two) Arrow-Debreu securities defined by

$$W^\uparrow : W^\uparrow(1,\uparrow) = 1, \quad W^\uparrow(1,\downarrow) = 0 \qquad (3.8a)$$
$$W^\downarrow : W^\downarrow(1,\uparrow) = 0, \quad W^\downarrow(1,\downarrow) = 1. \qquad (3.8b)$$

Write $W^\uparrow(0)$ (resp., $W^\downarrow(0)$) for the Arrow-Debreu prices at time $t = 0$. Then a general claim W can be written

$$W(1) = W(1,\uparrow) W^\uparrow(1) + W(1,\downarrow) W^\downarrow(1).$$

This implies by Theorem 1.2 that

$$W(0) = W(1,\uparrow)W^\uparrow(0) + W(1,\downarrow)W^\downarrow(0),$$

which will be the same as equation (3.4), as we shall see.

Of course, type 2 non arbitrage implies $W^\uparrow(0) > 0$ and $W^\downarrow(0) > 0$. $W^\uparrow(0)$ and $W^\downarrow(0)$ are called **state prices**. Arrow-Debreu securities act as a basis for the payoffs of any other security at $t = 1$.

For $W^\uparrow$:

$$1 = H_0 R_d + H_1 R_f X(1,\uparrow)$$
$$0 = H_0 R_d + H_1 R_f X(1,\downarrow)$$

implies

$$H_0 = \frac{-X(1,\downarrow)}{R_d \left[X(1,\uparrow) - X(1,\downarrow)\right]}$$
$$H_1 = \frac{1}{R_f \left[X(1,\uparrow) - X(1,\downarrow)\right]},$$

so

$$W^\uparrow(0) = H_0 + H_1 X(0)$$
$$= \frac{-\frac{1}{R_d}X(1,\uparrow) + \frac{1}{R_f}X(0)}{[X(1,\uparrow) - X(1,\downarrow)]}$$
$$= \frac{1}{R_d} \frac{\frac{R_d}{R_f}X(0) - X(1,\downarrow)}{X(1,\uparrow) - X(1,\downarrow)}$$
$$= \frac{\pi}{R_d}.$$

As $X(1,\downarrow) < X(1,\uparrow)$ we must also have $\frac{R_d}{R_f}X(0) > X(1,\downarrow)$.

For $W^\downarrow$, we proceed in the same way:

$$0 = H_0 R_d + H_1 R_f X(1,\uparrow)$$
$$1 = H_0 R_d + H_1 R_f X(1,\downarrow)$$

implies

$$H_0 = \frac{X(1,\uparrow)}{R_d[X(1,\uparrow) - X(1,\downarrow)]}$$
$$H_1 = \frac{-1}{R_f[X(1,\uparrow) - X(1,\downarrow)]},$$

so

$$W^{\downarrow}(0) = H_0 + H_1 X(0)$$
$$= \frac{\frac{1}{R_d}X(1,\uparrow) - \frac{1}{R_f}X(0)}{[X(1,\uparrow) - X(1,\downarrow)]}$$
$$= \frac{1}{R_d}\frac{X(1,\uparrow) - \frac{R_d}{R_f}X(0)}{X(1,\uparrow) - X(1,\downarrow)}$$
$$= \frac{1-\pi}{R_d}.$$

As $X(1,\downarrow) < X(1,\uparrow)$ we must also have $\frac{R_d}{R_f}X(0) < X(1,\uparrow)$.

We now discuss an exchange rate contingent claim using the CRR notation. With

$$X(1,\uparrow) = uX(0)$$
$$X(1,\downarrow) = dX(0)$$

we must have

$$0 < d < \frac{R_d}{R_f} < u.$$

Then

$$\pi = \frac{\frac{R_d}{R_f} - d}{u - d}$$

and

$$1 - \pi = \frac{u - \frac{R_d}{R_f}}{u - d}.$$

Remark 3.4. In the financial press some exchange rates are quoted in the European way and some in the American way. This can be confusing. The best way to resolve this is to obtain a list **world value of the US dollar** which writes 1 USD in terms of all the world currencies. This is published each Monday in the *Wall Street Journal*. Then all confusion can be removed. The web page www.oanda.com/converter/classic has details for 164 currencies from about 1990 to the present.

Examples of FX Derivatives

Example 3.5 (European call options).

We will discuss options on the USD.

These are traded **over the counter** (OTC) and on some exchanges (for example, on the Philadelphia Stock Exchange see http://www.phlx.com; see also Hull [37]).

We wish to discuss a contract that gives the holder the right (but not the obligation) to buy F USD at a fixed exchange rate K at time T. We need to specify the amount (face value) F, the exchange rate K and the expiry date T. At time T, the value $C(T)$ of this call option is given by

$$C(T) = \begin{cases} F[X(T) - K] & \text{if } X(T) > K \\ 0 & \text{if } X(T) \leq K, \end{cases}$$

where $X(T)$ is the (directly quoted) USD/CAD exchange rate at time T. If $X(T) > K$ we can buy F USD for $F \cdot K$ CAD, and we can sell the F USD for $F \cdot X(T)$ CAD, making a profit of $F \cdot X(T) - F \cdot K = F[X(T) - K]$. If $X(T) \leq K$ we let the option lapse, as it is cheaper to buy the USD in the market at rate $X(T)$.

We can also write

$$C(T) = F[X(T) - K]^+. \tag{3.9}$$

It is convenient to use face value for the call, $F = 1$. The premium of this call option is then multiplied by the face value to obtain the price of calls with various face values F. When $F = 1$, the price is often quoted as a percentage of the face value. So if the quote is 5% and the face value is 100,000 USD, then the premium is $0.05 \times 100{,}000 = 5000$ CAD (note the change in units). In the one-step binomial asset pricing model we obtain

$$C(0) = \frac{\pi}{R_d}[uX(0) - K]F. \tag{3.10}$$

Here we have used the CRR notation and we have assumed that

$$X(1, \downarrow) < K < X(1, \uparrow).$$

Note that the value of this call depends on the following inputs: u, d, R_d, R_f, K, F, $X(0)$.

Example 3.6 (An application of an FX call option). Suppose a Canadian company, ABC Inc., puts in a bid F for a USD denominated asset, (for example, a **warehouse** to hold exports before sales to the US market), and the success

50   3 The Binomial Model for Other Contracts

or failure of the bid will not be known until time T. As of now, you will not know the CAD amount that will have to be paid (it will be the uncertain amount $F \cdot X(T)$). To **cap** costs, ABC can buy a European (style) call option with strike rate K and face value F to expire at time T. Then you need to pay at most $F \cdot K$ CAD, as you will exercise the call at T if $X(T) > K$ to buy the F USD for $F \cdot K$ CAD.

Example 3.7 (European put option). This is just the same as the European Call Option except that the right to buy is replace by the right to sell. The value $P(T)$ of the put at expiry T is given by

$$P(T) = F\left[K - X(T)\right]^+, \qquad (3.11)$$

where (again) the face value is F and the strike rate is K. If $K > X(T)$ you can buy F USD for $F \cdot X(T)$ and exercise the put to sell the F USD for $F \cdot K$ to yield a profit $F[K - X(T)]$. If, on the other hand, $K \leq X(T)$, then let the option lapse, and sell the F USD at the higher market rate $X(T)$.

Example 3.8 (An application of an FX put option). A Canadian export company, EXP Inc., sells goods in the United States. Every time EXP issues bills for the export goods, it can take up to 3 months to receive payment (e.g., 100 million USD). For 3 months time EXP will not know the CAD value of the payments. To overcome this currency-risk problem, EXP could take out a 3-month put option on 100 million USD ($= F$) at exchange rate K (with premium passed onto the clients, no doubt!). Then EXP will be guaranteed at least $100 \cdot K$ million CAD for the exports.

Example 3.9 (Forward exchange rate contract). Again we have **long** and **short** contract, depending on whether one is **buying** or **selling**.

If you hold a **long forward exchange rate contract**, then you have the **obligation to buy** a specified amount F USD at a specified exchange rate K at time T, no matter what the actual exchange rate is at time T. No payment is made initially, and the profit at time T is $F\left[X(T) - K\right]$, which can be negative. If the present value of this profit is zero, then K is called the **T-forward exchange rate**. The time $t = 0$ value of the forward contract is 0 so this forward rate K is given by

$$K = \frac{R_d}{R_f} X(0). \qquad (3.12)$$

Both model-dependent and model-independent proofs can be given for (3.12).

Proof (For the one-step binomial asset pricing model). Using equation (3.4), we have

3.3 Exchange Rates

$$0 = \frac{1}{R_d}\left[\pi(F \cdot X(1,\uparrow) - F \cdot K) + (1-\pi)(F \cdot X(1,\downarrow) - F \cdot K)\right]$$

$$= \frac{1}{R_d}\left[\pi X(1,\uparrow) + (1-\pi)X(1,\downarrow)\right] \cdot F - \frac{K}{R_d} \cdot F$$

$$= \left[\frac{X(0)}{R_f} - \frac{K}{R_d}\right] \cdot F.$$

Here we used the identity

$$X(0)\frac{R_d}{R_f} = \pi X(1,\uparrow) + (1-\pi)X(1,\downarrow) \qquad (3.13)$$

since

$$\pi X(1,\uparrow) + (1-\pi)X(1,\downarrow) = \frac{\left(\frac{R_d}{R_f}X(0) - X(1,\downarrow)\right)X(1,\uparrow)}{X(1,\uparrow) - X(1,\downarrow)}$$
$$+ \frac{\left(X(1,\uparrow) - \frac{R_d}{R_f}X(0)\right)X(1,\downarrow)}{X(1,\uparrow) - X(1,\downarrow)}$$
$$= \frac{R_d}{R_f}X(0)$$

as required. □

Remark 3.10. Formula (3.12) is called the **exchange rate (covered) interest rate parity formula**. Often one sets

$$R_d = e^{r_d \Delta t}$$
$$R_f = e^{r_f \Delta t}$$

where Δt is the time interval (measured in years) from $t = 0$ to $t = 1$ ($T = 1$). In that case

$$K = e^{(r_d - r_f)\Delta t}X(0). \qquad (3.14)$$

There are many academic articles discussing or testing the validity of this formula. For example, consult Anthony [2] and Jacque [40].

If $r_d > r_f$, then we might expect the exchange rate to increase (the CAD becomes weaker with respect to USD). If this did not happen, then a U.S.-based investor could borrow USD in the U.S. at interest rate r_f, buy CAD and invest in CAD at interest rate r_d, convert back, and obtain a riskless profit. The **(covered) interest rate parity formula** shows a relationship between exchange rates and interest rates in the two countries.

Proof (A model-independent proof).

(a) Suppose (if possible) that $K > \frac{R_d}{R_f} X(0)$, where K is the T-forward rate.

At time $t = 0$, borrow $X(0)/R_f$ CAD, put $1/R_f$ USD (equal in value to $X(0)/R_f$ CAD) in a U.S. bank. Enter a T-forward contract to sell 1 USD for K CAD at time T. The net position is

$$\frac{X(0)}{R_f} - \frac{1}{R_f} \cdot X(0) + 0 = 0 \text{ CAD.}$$

At time $t = T$, repay the loan, realize 1 USD in your U.S. bank account, use the forward contract to convert to K CAD. The net position is

$$-\frac{X(0)}{R_f} \cdot R_d + K > 0 \text{ CAD}$$

and so we have a certain profit at time $t = T$ with no net outlay at time $t = 0$. This is an arbitrage opportunity in contradiction to the axiom.

(b) Also, $K < \frac{R_d}{R_f} X(0)$ leads to a similar contradiction.

At time $t = 0$, borrow $1/R_f$ USD, convert to CAD and invest in a Canadian bank. Enter a T-forward contract to buy 1 USD for K CAD at time T. The net position is

$$-\frac{X(0)}{R_f} + \frac{1}{R_f} \cdot X(0) + 0 = 0 \text{ CAD.}$$

At time $t = T$, realise the bank investment, buy 1 USD for K CAD by exercising the forward contract, repay the USD loan. The net position is

$$\frac{X(0)}{R_f} \cdot R_d - K > 0 \text{ CAD.}$$

Again there is a certain profit at time $t = T$ with no net outlay at time $t = 0$. This is again an arbitrage opportunity in contradiction to the axiom. □

Thus we have proved that the assumption of non arbitrage (together with other assumptions about no transaction costs, etc.) implies the **(covered) interest rate parity formula**.

Remark 3.11. Later we shall treat **exchange rate futures contracts**. For this we will need to introduce the concept of a **margin account**.

Example 3.12 (Calls in CAD equal puts in USD). Suppose you have a European call option to buy F USD using exchange rate K at time T. Then you

3.3 Exchange Rates

have the right to give up $F \cdot K$ in exchange for F USD. This is the **view** from Canada.

From the U.S. point of view, this is giving up $F \cdot K$ CAD for F USD, so we hold a European put option. In addition from the U.S. point of view the face value is $F \cdot K$ CAD, and the strike rate is $1/K$.

Let us set $T = 1$ and analyze this situation with the one-step binomial asset pricing model.

From Canada

From above,

$$C(0) = \frac{\pi_d}{R_d}[uX(0) - K] \cdot F$$

with

$$\pi_d = \frac{\frac{R_d}{R_f} - d}{u - d}.$$

From United States

We need to develop the U.S.-based one-step binomial asset pricing formula. Now all is denominated in USD, which is regarded as the domestic currency. Let $Y(t) = 1/X(t)$; then

$$Y(1,\uparrow) = \widehat{u}\, Y(0)$$
$$Y(1,\downarrow) = \widehat{d}\, Y(0)$$
$$\widehat{u} = \frac{1}{u}$$
$$\widehat{d} = \frac{1}{d}$$
$$\pi_f = \frac{\frac{R_f}{R_d} - \widehat{d}}{\widehat{u} - \widehat{d}}$$
$$\widehat{W}(0) = \frac{1}{R_f}\left[\pi_f \widehat{W}(1,\uparrow) + (1 - \pi_f)\widehat{W}(1,\downarrow)\right] \qquad (3.15)$$

and everything is expressed in USD. So $\widehat{W}(0)$ is the value in USD of an asset whose values at $t = 1$ are $\widehat{W}(1,\uparrow)$ in $\uparrow$ and $\widehat{W}(1,\downarrow)$ in $\downarrow$.

From the U.S. point of view the call was a put: the right to sell $F \cdot K$ CAD, which is foreign currency from the U.S. point of view, for F (that is, to receive F USD). The exchange rate is $\frac{1}{K}$ from the U.S. point of view. This is a put. So

$$\widehat{W}(1) = F \cdot K \left[\frac{1}{K} - Y(1)\right]^+ \tag{3.16}$$

which is the same as $F[X(1) - K]^+$ expressed in USD. In fact it is easy to show that

$$\frac{F[X(1) - K]^+}{X(1)} = F \cdot K \left[\frac{1}{K} - Y(1)\right]^+.$$

Then

$$\widehat{W}(0) = \widehat{P}(0)$$
$$= \frac{\pi_f}{R_f} \cdot F \cdot K \cdot \left[\frac{1}{K} - \widehat{u}Y(0)\right].$$

We have assumed that $X(1,\downarrow) < K < X(1,\uparrow)$ so $Y(1,\uparrow) < \frac{1}{K} < Y(1,\downarrow)$. Note that $Y(1,\uparrow) < Y(1,\downarrow)$.

We now claim that

$$\frac{\pi_f}{R_f} \cdot F \cdot K \cdot \left[\frac{1}{K} - \widehat{u}Y(0)\right] \cdot X(0) = \frac{\pi_d}{R_d} [uX(0) - K] \cdot F.$$

Both sides of this equation are denominated in CAD. Now

$$\frac{\pi_f}{R_f} \cdot F \cdot K \cdot \left[\frac{1}{K} - \widehat{u}Y(0)\right] \cdot X(0) = \frac{\pi_f}{R_f} \cdot F \cdot [X(0) - \widehat{u}K]$$
$$= \frac{\pi_f}{R_f} \cdot \frac{1}{u} \cdot F \cdot [uX(0) - K].$$

We need only show that

$$\frac{\pi_f}{uR_f} = \frac{\pi_d}{R_d}. \tag{3.17}$$

In fact

$$\frac{\pi_f}{uR_f} = \frac{\frac{R_f}{R_d} - \frac{1}{d}}{\frac{1}{u} - \frac{1}{d}} \cdot \frac{1}{uR_f}$$
$$= \frac{dR_f - R_d}{R_d \cdot d} \cdot \frac{ud}{d-u} \cdot \frac{1}{u \cdot R_f}$$

$$= \frac{dR_f - R_d}{d - u} \cdot \frac{1}{R_f} \cdot \frac{1}{R_d}$$

$$= \frac{\frac{R_d}{R_f} - d}{u - d} \cdot \frac{1}{R_d}$$

$$= \frac{\pi_d}{R_d},$$

so

$$\pi_f = u \cdot \frac{R_f}{R_d} \cdot \pi_d \qquad (3.18)$$

and also

$$1 - \pi_f = d \cdot \frac{R_f}{R_d} \cdot (1 - \pi_d). \qquad (3.19)$$

An interesting observation is

$$C(0) = \frac{\pi_d}{R_d} [uX(0) - K] \cdot F$$

$$= \left[\pi_f \cdot \frac{X(0)}{R_f} - \pi_d \cdot \frac{K}{R_d} \right] \cdot F, \qquad (3.20)$$

and because the call (from Canada) is equivalent to a put (from U.S.) the contract is also called USD PUT/CAD CALL to indicate that the contract can be viewed either way (and priced in either country).

3.4 Interest Rate Derivatives

Along with FX-derivatives, **interest rate derivatives** are the most commonly traded. Together they make up probably more than 85% of all derivatives traded.

An important function that we shall use is

$$P(t, T),$$

which indicates the value at time t of 1 CAD at expiry time T. If there are two or more currencies then we write

$$P^d(t, T)$$

for domestic and

$$P^f(t,T) \qquad (3.21)$$

for foreign (or $P^{f_1}(t,T), P^{f_2}(t,T), \ldots$, if necessary). The quantity $P(t,T)$ will be (usually) **unknown** until time t. It is the value at time t of a **zero-coupon bond** expiring at time T with (face) value 1 USD.

$P(t,T)$ is a very important function and can be used to compute the **present values** of **known** future cash flows. For example, if there are known (they are not random) cash flows

$$C_1, C_2, C_2, \ldots, C_m$$

at future times $t_1 < t_2 < t_3 < \ldots < t_m$, then the present (time $t = 0$) value of these cash flows is

$$\sum_{i=1}^{m} C_i \, P(0, t_i) \equiv C_1 \, P(0, t_1) + C_2 \, P(0, t_2) + \ldots + C_m \, P(0, t_m).$$

We now discuss a binomial asset pricing model for interest rates.

Consider two times $t = 0$ and $t = 1$.

For various times $T \geq 2$, there is a zero-coupon bond whose value at time $t = 0$ is $P(0, T)$. At the future time $t = 1$ in the binomial model there are two states of the world, $\uparrow$ and $\downarrow$. $P(t, T)$ will be uncertain and we suppose it can have two values,

$$P(1, T, \uparrow) \quad \text{and} \quad P(1, T, \downarrow).$$

Recall that $B(0) = 1$ and

$$B(1, \uparrow) = B(1, \downarrow) = R = \frac{1}{P(0,1)}.$$

The risk-neutral probabilities are

$$\pi = \frac{P(0,T)R - P(1,T,\downarrow)}{P(1,T,\uparrow) - P(1,T,\downarrow)} \qquad (3.22a)$$

$$1 - \pi = \frac{P(1,T,\uparrow) - P(0,T)R}{P(1,T,\uparrow) - P(1,T,\downarrow)}, \qquad (3.22b)$$

and the **one-step binomial asset pricing model** gives

3.4 Interest Rate Derivatives

$$W(0) = \frac{1}{R}[\pi W(1,\uparrow) + (1-\pi)W(1,\downarrow)] \quad (3.23)$$

as the present value of any interest rate contingent claim W.
In particular **for any other** $T \geq 2$ we have

$$P(0,T) = \frac{1}{R}[\pi P(1,T,\uparrow) + (1-\pi)P(1,T,\downarrow)]$$

which implies by rearrangement that, as in (3.22a),

$$\pi = \frac{P(0,T)R - P(1,T,\downarrow)}{P(1,T,\uparrow) - P(1,T,\downarrow)}.$$

This says that under no arbitrage, the definition of π in (3.22a) above, is independent of T. In other words, we must obtain the same value of π in (3.22a) no matter which value of $T \geq 2$ we choose.

This observation implies restrictions on the possible interest rate models in this framework. Consider quantities R, $R(1,\uparrow)$, and $R(1,\downarrow)$ defined as follows. If we invest (in a bank, under interest) \$1 at time $t=0$ then we will have \$$R$ at time $t=1$. Likewise, if we invest \$1 at time $t=1$ in $\uparrow$ then we will have \$$R(1,\uparrow)$ at time $t=2$. If we invest \$1 at time $t=1$ in $\downarrow$ then we will have \$$R(1,\downarrow)$ at time $t=2$.

Consequently,

$$P(0,1) = \frac{1}{R} \quad (3.24)$$

$$P(0,2) = \frac{1}{R}\left[\frac{\pi}{R(1,\uparrow)} + \frac{1-\pi}{R(1,\downarrow)}\right]. \quad (3.25)$$

This implies that, if we are given R and π, then we cannot select $R(1,\uparrow)$ and $R(1,\downarrow)$ in an arbitrary way, but there are still infinitely many choices in (3.25) for $R(1,\uparrow)$ and $R(1,\downarrow)$ under this restriction.

We now describe some choices.

Choice 1 (Ho and Lee)

Given π and k we assume that $R(1,\uparrow) = kR(1,\downarrow)$, then

$$P(0,2) = \frac{1}{R}\left[\frac{\pi}{R(1,\uparrow)} + \frac{(1-\pi)k}{R(1,\uparrow)}\right]$$

$$= \frac{1}{R}[\pi + (1-\pi)k] \cdot \frac{1}{R(1,\uparrow)},$$

58   3 The Binomial Model for Other Contracts

from which we solve for $R(1,\uparrow)$ and then $R(1,\downarrow) = \frac{1}{k}R(1,\uparrow)$.
If we write

$$R = 1 + r(0)$$
$$R(1,\uparrow) = 1 + r(1,\uparrow)$$
$$R(1,\downarrow) = 1 + r(1,\downarrow)$$

then we can deduce relationships between interest rates $r(1,\uparrow)$ and $r(1,\downarrow)$.
Some references for this model are Panjer et al. [59], Ho and Lee [33] and Pedersen et al. [60].

Choice 2 (Black, Derman and Toy)

Given π and $\sigma(1) > 1$, we assume that

$$r(1,\uparrow) = \sigma(1)r(1,\downarrow).$$

We then solve for $r(1,\downarrow)$ in

$$P(0,2) = \frac{1}{R}\left[\frac{\pi}{1+\sigma(1)r(1,\downarrow)} + \frac{1-\pi}{1+r(1,\downarrow)}\right]$$

and set $r(1,\uparrow) = \sigma(1)r(1,\downarrow)$.
References for this model include Panjer [59] and Black, Derman and Toy [7].

Remark 3.13. In Choice 1, the Ho and Lee model, k is a measure of volatility (spread) of the interest rates. The larger the value of k the larger the difference between $R(1,\uparrow)$ and $R(1,\downarrow)$ and, hence, the larger the difference between $r(1,\uparrow)$ and $r(1,\downarrow)$. The same comment can be made about $\sigma(1)$ in Choice 2.

Example 3.14 (Ho and Lee).
Given $P(0,1) = 0.95$, $P(0,2) = 0.90$, $\pi = 0.5$, $k = 1.04$, then

$$R = \frac{1}{0.95} = 1.05263...$$
$$r(0) = R - 1 = 0.05263... = 5.263\%$$
$$R(1,\uparrow) = \frac{\pi + (1-\pi)k}{R \cdot P(0,2)} = 1.07666...$$
$$r(1,\uparrow) = R(1,\uparrow) - 1 = 0.07666... = 7.666...\%$$
$$R(1,\downarrow) = \frac{R(1,\uparrow)}{1.04} = 1.035256...$$
$$r(1,\downarrow) = R(1,\downarrow) - 1 = 0.035256... = 3.5256...\%.$$

3.4 Interest Rate Derivatives

Example 3.15 (Black, Derman and Toy). Given $P(0,1) = 0.95$, $P(0,2) = 0.90$, $\pi = 0.5$, $\sigma(1) = 1.1$, we solve

$$0.90 = 0.95 \left[\frac{0.5}{1 + 1.1r(1,\downarrow)} + \frac{0.5}{1 + r(1,\downarrow)} \right] \tag{3.26}$$

to give

$$r(1,\downarrow) = 0.05292... = 5.292..\%$$
$$r(1,\uparrow) = \sigma(1)r(1,d) = 0.058212... = 5.8212...\%.$$

Here equation (3.26) can be solved for $r(1,\downarrow)$ in various ways: solution of quadratic formula, interval division method, Newton-Raphson method, etc.

Example 3.16 (Forward contracts). Consider a long forward contract agreeing to buy a zero-coupon bond with maturity T at time $t = 1$ for $\$K$ CAD. (We assume that $T \geq 2$). If this contract is entered by no payment of premium, then the present value of this contract is zero. Therefore, the rational price is

$$K = \frac{P(0,T)}{P(0,1)}. \tag{3.27}$$

This result can be shown in any model, so again this result is model-independent.

Proof (of equation 3.27). (a) Assume (if possible) that

$$\frac{P(0,T)}{P(0,1)} - K > 0. \tag{3.28}$$

At $t = 0$, short sell one T-zero, invest $K \cdot P(0,1)$ in a bank (or equivalently, buy K 1-zeros), enter a long forward to buy a T-zero at $t = 1$ for K CAD. The net proceeds at $t = 0$ are $P(0,T) - K \cdot P(0,1) + 0 > 0$ by (3.28). Pocket this profit.

At time $t = 1$, use the forward contract to purchase the zero-coupon bond and return it; this is funded from the investment K. There are no unfunded liabilities at time $t = 1$, so we have created an arbitrage. This shows that (3.28) must be false.

(b) Assume (if possible) that

$$K - \frac{P(0,T)}{P(0,1)} > 0. \tag{3.29}$$

At $t = 0$, borrow $K \cdot P(0,1)$, (or short sell K 1-zeros), purchase one T-zero, enter a short forward contract to sell one T-zero at time $t = 1$ for K CAD. The net proceeds at $t = 0$ are $-P(0,T) + K \cdot P(0,1) + 0 > 0$ by (3.29). Pocket this profit.

At time $t = 1$, use the forward contract to sell the T-zero for $\$K$, use this to repay the loan $\$K$. There are no unfunded liabilities at time $t = 1$, so we have created an arbitrage. This shows that (3.29) must be false.

From (a) and (b), we have established the result. □

Example 3.17 (European options). Consider the option to purchase a T-zero at time $t = 1$ for $\$K$, where $T \geq 2$. The value at time $t = 1$ is

$$C(1) = (P(1,T) - K)^+,$$

and if $P(1,T,\uparrow) < K < P(1,T,\downarrow)$, then the value is given as

$$C(0) = \frac{1-\pi}{R}[P(1,T,\downarrow) - K]$$

in the one-step binomial asset pricing model. Recall that there is an inverse relationship between bond prices and interest rates.

Likewise the corresponding European put option has present value

$$P(0) = \frac{\pi}{R}[K - P(1,T,\uparrow)].$$

Combining these call and put formulae:

$$C(0) - P(0) = P(0,T) - \frac{K}{R} = P(0,T) - KP(0,1). \qquad (3.30)$$

This equation can also be proved using a model-independent argument. Here are some of the details. If

$$C(0) - P(0) - P(0,T) + K \cdot P(0,1) > 0, \qquad (3.31)$$

then at time $t = 0$, short sell one call, buy a put, buy a T-zero, borrow $K \cdot P(0,1)$ (short sell K 1-zeros). Pocket the net profit.

At time $t = 1$, cash settle the call, realize the value of the put, sell the T-zero, repay the loan. The net of all of this is

$$-(P(1,T) - K)^+ + (K - P(1,T))^+ + P(1,T) - K = 0,$$

which can be seen by considering $P(1,T) > K$ and $P(1,T) \leq K$.

The reverse inequality in (3.31) can be established by a similar argument.

Example 3.18 (Numerical example). Let us suppose that $r(0) = 3\%$, $r(1,\uparrow) = 4\%$, $r(1,\downarrow) = 2\%$. Let us suppose that time interval from $t = 0$ to $t = 1$ is one month or $\frac{1}{12}$ years. We also know that $P(0,2) = 0.9950$, $K = 0.997$. We wish to find $C(0)$ if

$$C(1) = (P(1,2) - K)^+.$$

Now

$$R = 1 + \frac{0.03}{12} = 1.0025 = \frac{1}{P(0,1)}$$

$$P(1,2,\uparrow) = \frac{1}{1 + \frac{0.04}{12}} = 0.99667774\ldots$$

$$P(1,2,\downarrow) = \frac{1}{1 + \frac{0.02}{12}} = 0.99833610\ldots$$

$$\pi = \frac{P(0,2)R - P(1,2,\downarrow)}{P(1,2,\uparrow) - P(1,2,\downarrow)} = 0.511712131\ldots$$

$$C(1) = \begin{cases} 0, & \text{in } \uparrow \\ 0.001336106\ldots, & \text{in } \downarrow \end{cases}$$

$$C(0) = \frac{1-\pi}{R}[P(1,2,\downarrow) - K]$$

$$= 0.0006507777\ldots$$

$$= 0.065\%$$

$$= 6.5 \text{ basis points.}$$

If the face value were $1000 then $K = \$997$ and the value of the (call) option would be $\$0.00065 \times 1000 = \$0.65 = 65$ cents.

Remark 3.19. We shall return to interest rate concepts after we have discussed multi-period models.

Let us note an interesting difference between stock prices and interest rates. If we know a stock price at $t = 0$, then we cannot know its price with certainty at $t = 1$, but if we invest $1 at $t = 0$ then we know the value of this investment at $t = 1$ (though at $t = 0$ we will not know its value at $t = 2$). We sometimes say that interest rates are predictable processes: The interest payable at time t is known at time $t - 1$.

3.5 Exercises

Exercise 3.20. Suppose we have a market with two tradeable assets (as in the basic model of the previous chapter). For the risky asset we have $\tilde{S}$ with

$\tilde{S}(0) = 0$, $\tilde{S}(1,\uparrow) > 0$ and $\tilde{S}(1,\downarrow) < 0$. The riskless asset is the same as before. Now derive again (2.12) but using $\tilde{S}$ in place of S. Give an example for $\tilde{S}$.

Exercise 3.21 (Forward contract). Recall that a **forward contract** is an agreement to buy or sell an asset at a certain future time for a certain price (the delivery price). At the time the contract is entered into, the delivery price is chosen so that the value of the forward contract to both parties is zero. This means it costs nothing to take either a long or short position in a forward contract. If you hold the long position with the obligation to buy S at time T with delivery price K, draw a graph of your profit at time T. Explain how a forward contract can be used for either speculation or hedging. If you are bullish about S, you could take long positions in either call options or forward contracts. Discuss the similarities and differences, including a discussion of differences in risks taken with these two strategies.

Exercise 3.22. Consider again the one-period binomial asset pricing model. For a stock, let us write $S(0) = S$, $S(1,\uparrow) = S(\uparrow)$ and $S(1,\downarrow) = S(\downarrow)$. (Assume here that $S(\uparrow) > S(\downarrow)$). Let F be the $t = 1$ forward price of the stock. Let C be the ($t = 0$) value of the ATM call on the stock. Let (as usual) R be the value in CAD at $t = 1$ of 1 CAD at $t = 0$. Show the following:

1. $S(\downarrow) < F < S(\uparrow)$
2. $RC < F < RC + S$
3. Given the relationships

$$S^2 = S(\uparrow)S(\downarrow)$$
$$F = \pi S(\uparrow) + (1 - \pi)S(\downarrow)$$
$$C = \frac{\pi}{R}(S(\uparrow) - S)$$

find π, $S(\uparrow)$ and $S(\downarrow)$ in terms of S, R, F, C. Hint: Show that

$$S(\uparrow) = \frac{S[R\ C + S]}{F - R\ C} \tag{3.32}$$

and that the right hand side of (3.32) exceeds F. This leads to an alternative calibration of the the binomial asset pricing model.

Remark 3.23. This is the basis of an idea used by Derman and Kani of Goldman-Sachs (NY) in 1994 [24]. The idea is that S, R, F, C are market information that will allow us to compute the possible values $S(\uparrow)$ and $S(\downarrow)$.

Exercise 3.24. Study Examples 3.5–3.8.

3.5 Exercises

For Exercises 3.25 and 3.26 you should use the following information.

The two currency problems below are modelled in a one-step binomial asset pricing model.

$X(0)$, $X(1,\uparrow)$ and $X(1,\downarrow)$ will refer to USD/CAD exchange rates (directly quoted), and $J(0)$, $J(1,\uparrow)$ and $J(1,\downarrow)$ will refer to JPY/CAD exchange rates (directly quoted). JPY is the abbreviation for the Japanese yen.

Write R_d (resp., R_a, R_j), for the values at $t = 1$ in CAD (resp., USD, JPY), of one CAD (USD, JPY). The 30-day annual interest rates for the various currencies are 5.12%, 5.74% and 0.65%, respectively. So $R_d = 1 + \frac{0.0512}{12} \approx 1.004267$, $R_a \approx 1.004783$ and $R_j \approx 1.000542$.

Suppose $X(0)$ and $J(0)$ were 1.2195 (about 82 cents), and 0.011377 (about 88 JPY to CAD). Take $X(1,\downarrow) = 1.2821$ and $J(1,\downarrow) = 0.012599$. We shall take $X(1,\uparrow) = 1.1905$, which corresponds to 84 cents. We shall also use the notation

$$X(1,\uparrow) = u_a X(0), \quad X(1,\downarrow) = d_a X(0), \quad J(1,\uparrow) = u_j J(0), \quad J(1,\downarrow) = d_j J(0).$$

Exercise 3.25 (Call-put parity for currency options). Let $P(0)$ (resp., $C(0)$) be the present ($t = 0$) value of a put (call) option to sell (resp., buy) 1 USD for K CAD (exchange rate K) at time $t = 1$. If the current exchange rate is $X(0)$, show that

$$C(0) - P(0) = \frac{X(0)}{R_a} - \frac{K}{R_d}. \tag{3.33}$$

You can use either a model-independent argument or the appropriate pricing formulae from the one-step binomial asset pricing model.

Find the value (at $t = 0$) of the call option to buy 1 USD at time $t = 1$ for 1.2195 CAD. For this you will need the appropriate pricing formula

$$f(0) = \frac{\pi f(1,\uparrow) + (1-\pi) f(1,\downarrow)}{R_d} \tag{3.34}$$

where

$$\pi \equiv \pi_a = \frac{\frac{R_d}{R_a} - d_a}{u_a - d_a}. \tag{3.35}$$

Use equation (3.34) to price the corresponding put option to sell 1 USD at time $t = 1$ for 1.2195 CAD. When you have done this, verify the formula (3.33).

3 The Binomial Model for Other Contracts

Exercise 3.26. 1. Let π_a (the American π) be given by (3.35), and π_j (the Japanese π) be given the same formula but with all a replaced by j. Define the two Arrow-Debreu securities for the up and down states at $t = 1$. Give the price at $t = 0$ of the up Arrow-Debreu security using (a) the American exchange rates and (b) the Japanese exchange rates. Deduce that $\pi_a = \pi_j$.

2. Use 1. to show that $J(\uparrow) \approx 0.011111$.

3. Use (assume valid) the interest parity formula to compute the $t = 1$ forward exchange rates for USD/CAD, JPY/CAD and USD/JPY.

4. Canadian Corporation (CC) owns factories in both Japan and in the United States. Today ($t = 0$) the treasurer of CC goes to merchant bank XYZ to buy an option giving her the right to sell 10 million JPY for 93,688 USD at $t = 1$. By this she hopes to transfer cash from Japanese operations to the U.S. at about the forward rate. Show that the value of this option at $t = 1$ is (in CAD)

$$[-10,000,000J(1) + 93,688X(1)]^+.$$

Find the value of this option in CAD at $t = 0$.

4

Multiperiod Binomial Models

There are basically two types of trees that we shall consider:

- **recombining trees**, and
- **non-recombining trees**.

The latter will be used when we discuss **path-dependent options**. For now we shall consider only recombining trees, though the general methodology will be the same for both.

4.1 The Labelling of the Nodes

Each node in the tree has a label (n, j). The label n stands for time $n = 0, 1, 2, 3, \ldots$, and j numbers the states $j = 0, 1, 2, \ldots, n$. At times $t = n$ there are thus $n + 1$ states. If you are at node (n, j) at time $t = n$ then at time $t = n + 1$ one may be in either $(n + 1, j + 1)$ or $(n + 1, j)$.

Here we have generalized the one-step model notation. If we are at time $t = 0$ in $(0, 0)$ then at time $t = 1$ we can be in $(1, \uparrow)$ or $(1, \downarrow)$, which we now label as $(1, 1)$ or $(1, 0)$. From the point of view of node (n, j), node $(n+1, j+1)$ is $\uparrow$ and $(n + 1, j)$ is $\downarrow$.

An interpretation of node (n, j) is the following. We are at time $t = n$ and we moved there with j up-movements to get to this node. With recombining trees, there are many ways to reach node (n, j).

4.2 The Labelling of the Processes

In state (n, j) the stock price, exchange rate, etc. will be denoted by $S(n, j)$, $X(n, j)$ and so on.

4 Multiperiod Binomial Models

By $P_j^{(n)}(T)$ we will mean the value at time $t = n$ in state j of \$1 at time $t = T + n$ (T periods after $t = n$).

Let $R(n, j) \equiv 1 + r(n, j)$ be the value at time $t = n + 1$ of \$1 invested in a bank at time $t = n$ in (n, j) and held until time $t = n+1$. The quantity $r(n, j)$ is the interest that can be received at time $t = n+1$ starting at time $t = n$ in (n, j).

We shall also let $\lambda(n, j)$ denote the value at time $t = 0$ of the Arrow-Debreu security that pays \$1 at time $t = n$ in state j and \$0 (zero) at time $t = n$ in any other state.

For $m \leq n$ we shall later also use the notation $A(m, i, n, j)$ to denote the value at time $t = m$ in state i of the (Arrow-Debreu) security that pays at time $t = n$, \$1 in state j and nothing in the other states at $t = n$. Of course, this notation generalizes λ and clearly $\lambda(n, j) = A(0, 0, n, j)$.

4.3 Generalized Quantities

For Stocks

We define

$$u(n, j) = \frac{S(n+1, j+1)}{S(n, j)} \quad (4.1a)$$

$$d(n, j) = \frac{S(n+1, j)}{S(n, j)} \quad (4.1b)$$

$$\pi(n, j) = \frac{R(n, j) - d(n, j)}{u(n, j) - d(n, j)}. \quad (4.1c)$$

To avoid arbitrage, restrictions must be placed on the multiperiod model. This requires that

$$0 < \pi(n, j) < 1$$

hold for all (n, j). Thus, restrictions on $R(n, j)$, $u(n, j)$ and $d(n, j)$ are required by (4.1c), or equivalently on the $S(n, j)$.

For Currencies

Similar multiperiod quantities can be written for currencies, where we use X in place of S and the definition of $\pi(n, j)$ involves a $R_d(n, j)$ and an $R_f(n, j)$, so

$$u(n,j) = \frac{X(n+1,j+1)}{X(n,j)} \quad (4.2a)$$

$$d(n,j) = \frac{X(n+1,j)}{X(n,j)} \quad (4.2b)$$

$$\pi(n,j) = \frac{\frac{R_d(n,j)}{R_f(n,j)} - d(n,j)}{u(n,j) - d(n,j)}. \quad (4.2c)$$

Again restrictions must apply so that $0 < \pi(n,j) < 1$.

For Bonds

We may choose (any) $T \geq 2$ and define

$$\pi(n,j) = \frac{P_j^n(T) \cdot R(n,j) - P_j^{n+1}(T-1)}{P_{j+1}^{n+1}(T-1) - P_j^{n+1}(T-1)} \quad (4.3)$$

which is a generalization of (3.22a). Again, restrictions must apply so that $0 < \pi(n,j) < 1$.

4.4 Generalized Backward Induction Pricing Formula

In any application (with the appropriate definitions of π) we have the **backward induction pricing formula**.

$$W(n,j) = \frac{1}{R(n,j)} \left[\pi(n,j) W(n+1,j+1) + (1 - \pi(n,j)) W(n+1,j)\right]. \quad (4.4)$$

Also note the formula

$$W(0,0) = \sum_{j=0}^{N} \lambda(N,j) W(N,j). \quad (4.5)$$

This expresses the value of W at time $t = 0$ in terms of the possible values of W at time $t = N$. Later we shall give **Jamshidian's forward induction formula** [41] for the computation of the values of $\lambda(n,j)$. Of course $\lambda(1,0) = W^{\downarrow}(0)$ and $\lambda(1,1) = W^{\uparrow}(0)$, which we used earlier.

4.5 Pricing European Style Contingent Claims

In this case the

$$W(N, j) \text{ for } j = 0, 1, 2, \ldots, N$$

are given for some future time N. For example, in the case of the European call stock option expiring at time $t = N$ with strike price K, we have

$$W(N, j) = [S(N, j) - K]^+ \quad \text{for } j = 0, 1, 2, \ldots, N.$$

We can now use the backward induction formula (4.4) to compute $W(n, j)$ for each $n < N$ and $j = 0, 1, 2, \ldots, n$. In particular we can find $W(0, 0)$.

4.6 The CRR Multiperiod Model

The CRR model is basically the simplest **constant** model. Here we assume that there are real numbers

$$0 < d < R < u.$$

For all (n, j) we assume

$$u(n, j) = u \tag{4.6a}$$
$$d(n, j) = d \tag{4.6b}$$
$$R(n, j) = R \tag{4.6c}$$
$$\pi(n, j) = \pi = \frac{R - d}{u - d}. \tag{4.6d}$$

With $S(0, 0) = S$

$$S(n, j) = S(0, 0) u^j d^{n-j} \equiv S u^j d^{n-j}$$

and

$$W(n, j) = \frac{1}{R} \left[\pi W(n+1, j+1) + (1 - \pi) W(n+1, j) \right].$$

This recursion has an explicit solution,

$$W(0,0) = \frac{1}{R^N} \sum_{l=0}^{N} C_l^N \pi^l (1-\pi)^{N-l} W(N,l). \qquad (4.7)$$

For some intermediate time n, $0 \le n \le N$,

$$W(n,j) = \frac{1}{R^{N-n}} \sum_{l=0}^{N-n} C_l^{N-n} \pi^l (1-\pi)^{N-n-l} W(N, j+l) \qquad (4.8)$$

for $j = 0, 1, 2, \ldots, n$.

Here,

$$C_l^k \equiv \frac{k!}{l!(k-l)!} = \frac{k(k-1)(k-2)\cdots(k-l+1)}{l(l-1)(l-2)\cdots 3\cdot 2\cdot 1} \text{ for } 0 \le l \le k,$$

is one of the binomial coefficients. There are various proofs of (4.7) and (4.8). The CRR paper gives the usual one, which we now illustrate with $N = 2$. There is also a modern proof using Jamshidian's forward induction.

For the CRR proof

$$W(1,1) = \frac{\pi}{R} W(2,2) + \frac{(1-\pi)}{R} W(2,1)$$

$$W(1,0) = \frac{\pi}{R} W(2,1) + \frac{(1-\pi)}{R} W(2,0)$$

$$\begin{aligned} W(0,0) &= \frac{\pi}{R} W(1,1) + \frac{(1-\pi)}{R} W(1,0) \\ &= \frac{\pi}{R} \left[\frac{\pi}{R} W(2,2) + \frac{(1-\pi)}{R} W(2,1) \right] \\ &\quad + \frac{(1-\pi)}{R} \left[\frac{\pi}{R} W(2,1) + \frac{(1-\pi)}{R} W(2,0) \right] \\ &= \frac{1}{R^2} \left[\pi^2 W(2,2) + 2\pi(1-\pi) W(2,1) + (1-\pi)^2 W(2,0) \right] \end{aligned}$$

and so on. One can use the **principle of mathematical induction** to establish the general result (4.8).

4.7 Jamshidian's Forward Induction Formula

Recall $\lambda(n, j)$ is the value at $t = 0$ of the Arrow-Debreu security that pays \$1 at $t = n$ in state j and \$0 in any other state. We now obtain a forward induction formula to compute $\lambda(n, j)$ for each (n, j), the so called **state prices**.

4 Multiperiod Binomial Models

We start with

$$\lambda(0,0) = 1. \qquad (4.9)$$

For convenience we set $\lambda(n,j) = 0$ whenever $j < 0$ or $j > n$. Then

$$\lambda(n,j) = \frac{1 - \pi(n-1,j)}{R(n-1,j)}\lambda(n-1,j) + \frac{\pi(n-1,j-1)}{R(n-1,j-1)}\lambda(n-1,j-1) \qquad (4.10)$$

since the Arrow-Debreu security that pays \$1 at (n,j) is the same as (see (4.4)) the security that has payoff at $t = n - 1$

$$W(n-1,k) = \begin{cases} 0 & \text{if } k > j \\ \dfrac{1 - \pi(n-1,j)}{R(n-1,j)} & \text{if } k = j \\ \dfrac{\pi(n-1,j-1)}{R(n-1,j-1)} & \text{if } k = j - 1 \\ 0 & \text{if } k < j - 1. \end{cases} \qquad (4.11)$$

in the case that $1 \le j \le n - 1$. The result now follows from (4.5) with the choice $N = n - 1$.

In the case $j = 0$ we replace (4.11) by

$$W(n-1,k) = \begin{cases} 0 & \text{if } k > 0 \\ \dfrac{1 - \pi(n-1,0)}{R(n-1,0)} & \text{if } k = 0. \end{cases} \qquad (4.12)$$

In the case $j = n$ we replace (4.11) by

$$W(n-1,k) = \begin{cases} \dfrac{\pi(n-1,n-1)}{R(n-1,n-1)} & \text{if } k = n - 1 \\ 0 & \text{if } k < n. \end{cases} \qquad (4.13)$$

and use (4.5) as before with $N = n - 1$.

Remark 4.1. Another proof makes use of the observation that

$$W(0,0) = \sum_{j=0}^{n} \lambda(n,j) W(n,j)$$

$$= \sum_{k=0}^{n-1} \lambda(n-1,k) W(n-1,k)$$

$$= \sum_{k=0}^{n-1} \frac{\lambda(n-1,k)}{R(n-1,k)} \left[\pi(n-1,k)W(n,k+1) + (1-\pi(n-1,k)W(n,k)\right].$$

We now compare the coefficient of $W(n,j)$ in the first and last lines.

4.8 Application to CRR Model

The Jamshidian forward induction formula is (with $\lambda(0,0) = 1$ and $\pi = \frac{R-d}{u-d}$)

$$\lambda(n,j) = \begin{cases} \dfrac{\pi}{R} \lambda(n-1, n-1) & \text{if } j = n \\[6pt] \dfrac{1-\pi}{R} \lambda(n-1, j) + \dfrac{\pi}{R} \lambda(n-1, j-1) & \text{if } 1 \le j < n \\[6pt] \dfrac{1-\pi}{R} \lambda(n-1, 0) & \text{if } j = 0. \end{cases} \quad (4.14)$$

We now find the solution of (4.14). It is provided by the following lemma.

Lemma 4.2.
$$\lambda(n,j) = C_j^n \pi^j (1-\pi)^{n-j} R^{-n}, \tag{4.15}$$

where $C_j^n = \frac{n!}{j!(n-j)!}$ is the binomial coefficient. (For a discussion of binomial coefficients see Appendix A).

Proof. We shall prove (4.15) by the principle of mathematical induction. The formula (4.15) clearly holds when $n = 0$ (and $j = 0$).

Now assume that (4.15) holds when $n = k \ge 0$ and for $j = 0, 1, 2, \ldots, k$.
Now let $n = k+1$ and $1 \le j \le k$; then

$$\begin{aligned}
\lambda(k+1, j) &= \frac{1-\pi}{R} \lambda(k, j) + \frac{\pi}{R} \lambda(k, j-1) \\
&= \frac{1-\pi}{R} \left[\frac{C_j^k \pi^j (1-\pi)^{k-j}}{R^k}\right] \\
&\quad + \frac{\pi}{R} \left[\frac{C_{j-1}^k \pi^{j-1}(1-\pi)^{k-j+1}}{R^k}\right] \\
&= \left[C_j^k + C_{j-1}^k\right] \pi^j (1-\pi)^{k-j+1} R^{-k-1} \\
&= C_j^{k+1} \pi^j (1-\pi)^{k-j+1} R^{-k-1}.
\end{aligned}$$

So (4.15) also holds for $n = k+1$ and $j = 1, 2, 3, \ldots, k$. We must check it holds when $j = 0$ and $j = k+1$. In fact

$$\begin{aligned}
\lambda(k+1, 0) &= \frac{1-\pi}{R}\lambda(k, 0) \\
&= \frac{1-\pi}{R} C_0^k (1-\pi)^k R^{-k} \\
&= C_0^{k+1}(1-\pi)^{k+1} R^{-k-1}
\end{aligned}$$

since $C_0^{k+1} = C_0^k = 1$. Also

$$\begin{aligned}
\lambda(k+1, k+1) &= \frac{\pi}{R}\lambda(k, k) \\
&= \frac{\pi}{R} C_k^k \pi^k R^{-k} \\
&= C_{k+1}^{k+1} \pi^{k+1} R^{-k-1}
\end{aligned}$$

since $C_{k+1}^{k+1} = C_k^k = 1$.

In summary, (4.15) holds for $n = 0$ and for $n = k+1$ whenever it holds for $n = k \geq 0$. The principle of mathematical induction then implies that it holds for all integers $n = 0, 1, 2, \ldots$. □

Corollary 4.3. *In the CRR model we have*

$$W(0,0) = \frac{1}{R^N} \sum_{l=0}^{N} C_l^N \pi^l (1-\pi)^{N-l} W(N, l). \tag{4.16}$$

Proof. We use

$$W(0,0) = \sum_{l=0}^{N} \lambda(N, l) W(N, l)$$

and the formula for $\lambda(N, l) = C_l^N \pi^l (1-\pi)^{N-l} R^{-N}$.

Corollary 4.4. *In the CRR model we have*

$$W(n, j) = \frac{1}{R^{N-n}} \sum_{l=0}^{N-n} C_l^{N-n} \pi^l (1-\pi)^{N-n-l} W(N, j+l) \tag{4.17}$$

for $0 \leq n \leq N$ and $j = 0, 1, 2, \ldots, n$.

Proof. We can simply apply Corollary 4.3 by looking at the tree with vertices (n, j), (N, j), (N, N), and think of (n, j) as $(0, 0)$ and have an $N - n$ step tree. More formally, let $M = N - n$ and $V(k, l) \equiv W(k + n, l + j)$, then V satisfies the backwardization formula (4.4), and hence from Corollary 4.5,

$$V(0,0) = \sum_{l=0}^{M} \lambda(M,l) V(M,l)$$

and this is just (4.16). ∎

4.9 The CRR Option Pricing Formula

We now use this theory to write down the European call price in the multi-period CRR model. Recall

$$C(N,l) = [S(N,l) - K]^+ = \left[Su^l d^{N-l} - K\right]^+$$

with $S = S(0,0)$. We assume $Sd^N < K < Su^N$ and leave the other cases to the remarks below. Then

$$C(0,0) = \frac{1}{R^N} \sum_{l=0}^{N} C_l^N \pi^l (1-\pi)^{N-l} \left[Su^l d^{N-l} - K\right]^+$$

$$= \frac{1}{R^N} \sum_{l=a}^{N} C_l^N \pi^l (1-\pi)^{N-l} \left[Su^l d^{N-l} - K\right],$$

where a is the (unique) least integer so that

$$l \geq a \text{ implies } Su^l d^{N-l} > K$$
$$l < a \text{ implies } Su^l d^{N-l} \leq K.$$

As

$$Su^l d^{N-l} = Sd^N \left[\frac{u}{d}\right]^l$$

with $\frac{u}{d} > 1$, so $Su^l d^{N-l}$ is an increasing function of l. Therefore, a exists. We can write

$$C(0,0) = SJ_1 - \frac{K}{R^N} J_2$$

where

$$J_1 = \frac{1}{R^N} \sum_{l=a}^{N} C_l^N \pi^l (1-\pi)^{N-l} u^l d^{N-l}$$

$$= \sum_{l=a}^{N} C_l^N \left[\frac{\pi u}{R}\right]^l \left[\frac{(1-\pi)d}{R}\right]^{N-l}$$

and

$$J_2 = \sum_{l=a}^{N} C_l^N \pi^l (1-\pi)^{N-l}.$$

Note that

$$\frac{\pi u}{R} + \frac{(1-\pi)d}{R} = 1$$

so we can write

$$\pi' = \frac{\pi u}{R}, \quad 1 - \pi' = \frac{(1-\pi)d}{R}. \tag{4.18}$$

It is now convenient to make a definition.

Definition 4.5. *For $0 < p < 1$, set*

$$\Phi[a; n, p] = \sum_{l=a}^{n} C_l^n p^l (1-p)^{n-l} \tag{4.19}$$

Remark 4.6. $\Phi[a; n, p]$ is called the **complementary binomial distribution function**. We have

$$\Phi[0; n, p] = 1 \tag{4.20}$$

$$\Phi[a; n, p] = 1 - \sum_{l=0}^{a-1} C_l^n p^l (1-p)^{n-l}$$

$$\equiv 1 - F_n(a), \tag{4.21}$$

where F_n is called the **binomial distribution function**.

Toss a (biased) coin n times. If the probability of a head is p, and the probability of a tail is $1 - p$, then the probability that the number of heads in n tosses is less than a is given by $F_n(a)$.

Theorem 4.7.
$$C(0,0) = S\Phi[a; N, \pi'] - KR^{-N}\Phi[a; N, \pi] \qquad (4.22)$$

where a is the least integer with $S(N,a) > K$, $\pi' = \frac{\pi u}{R}$.

Remark 4.8. Formula (4.22) is the CRR formula for value of a European call option.

If $K \geq Su^N$ then $C(0,0) = 0$.
If $K \leq Sd^N$ then $C(0,0) = S - KR^{-N}$, as $C(N,l) \equiv S(N,l) - K$ for all l.

4.10 Discussion of the CRR Formula

Point 1

We can also price the **European put options** via the call-put parity formula for European options or in a direct way. In fact

$$P(0,0) = KR^{-N}\Psi[a; N, \pi] - S\Psi[a; N, \pi'] \qquad (4.23)$$

where
$$\Psi[a; N, p] \equiv 1 - \Phi[a; N, p]. \qquad (4.24)$$

Point 2—What is π'?

Recall the formulae (4.18) for π' and $1 - \pi'$. These can be compared with the formulae (3.18) and (3.19) in Section 3.3 where we discussed change of numeraires (when using either CAD or USD as the domestic currency) in exchange rate models. In fact if $R_f = 1$, then formulas (3.18) and (3.19) are exactly the same as (4.18). If π is the risk-neutral probability (of upstate) when CAD is the numeraire, π' is the risk-neutral probability (of upstate) when stock price is the numeraire. The proof is exactly the same as in the exchange rate example.

For completeness we write out some details.

In the **one-step binomial model** let $\widetilde{B} = \frac{S}{S} \equiv 1$ and $\widetilde{S} = \frac{B}{S}$. Then the stock becomes the riskless asset, and the riskless asset becomes risky.

$\widetilde{B}(0) = 1$, $\widetilde{B}(1,\uparrow) = 1$, $\widetilde{B}(1,\downarrow) = 1$. Thus $\widetilde{R} = 1$. $\widetilde{S}(0) = \frac{1}{S(0)}$, $\widetilde{S}(1,\uparrow) = \frac{R}{u}\widetilde{S}(0) \equiv \widetilde{u}\widetilde{S}(0)$, $\widetilde{S}(1,\downarrow) = \frac{R}{d}\widetilde{S}(0) \equiv \widetilde{d}\widetilde{S}(0)$.

So under this numeraire, the risk-neutral (up-) probability is

$$\widetilde{\pi} = \frac{\widetilde{R} - \widetilde{d}}{\widetilde{u} - \widetilde{d}}$$

$$= \frac{1 - \frac{R}{d}}{\frac{R}{u} - \frac{R}{d}}$$

$$= \frac{u}{R}\left[\frac{R-d}{u-d}\right] = \frac{u}{R}\pi = \pi'$$

and also

$$1 - \widetilde{\pi} = \frac{d}{R}(1 - \pi) = 1 - \pi'$$

as claimed.

In the **multistep binomial model**, $B(n,j) = R^n$, and with stock price as numeraire we have $\widetilde{B}(n,j) = \frac{S(n,j)}{S(n,j)} \equiv 1$, $\widetilde{S}(n,j) = \frac{B(n,j)}{S(n,j)} \equiv \frac{R^n}{S(n,j)}$, and again with the CRR model we have $\widetilde{\pi}(n,j) \equiv \pi'$ for all (n,j).

Point 3—What is $\Phi[a; N, \pi]$?

If we define

$$P^\pi\left[S(N) = Su^l d^{N-l}\right] = C_l^N \pi^l (1-\pi)^{N-l}$$

then

$$\Phi[a; N, \pi] = P^\pi\left[S(N) > K\right]$$

and

$$\Phi[a; N, \pi'] = P^{\pi'}\left[S(N) > K\right].$$

Point 4—Black and Scholes Formula

Let us make the definitions

$$\Delta t = \frac{T}{N}$$

4.10 Discussion of the CRR Formula

$$R = \exp(r\Delta t)$$

$$u = \exp(\sigma\sqrt{\Delta t})$$

$$d = \exp(-\sigma\sqrt{\Delta t})$$

where r is the risk-free interest rate that applies over the interval $[0, T]$ for continuous compounding. (If interest r is accumulated over a year and the interest is added and compounded every n-th part of the year then \$1 becomes $\$(1 + \frac{r}{n})^n$. In the limit this is e^r, and over a time period T, \$1 "continuously compounded" becomes $\$e^{rT}$.) The constant $\sigma > 0$ is called the **volatility**. We then let $N \to \infty$ in the CRR model. It can then be shown that

$$C(0,0) \to S\mathcal{N}(d_1) - Ke^{-rT}\mathcal{N}(d_2) \qquad (4.25)$$

where

$$\mathcal{N}(x) = \frac{1}{\sqrt{2\pi}} \int_{-\infty}^{x} e^{-\frac{1}{2}y^2} dy$$

and

$$e^{-rT} = R^{-N}$$

$$d_1 = \frac{\ln\left(\frac{S}{K}\right) + (r + \frac{1}{2}\sigma^2)T}{\sigma\sqrt{T}}$$

$$d_2 = \frac{\ln\left(\frac{S}{K}\right) + (r - \frac{1}{2}\sigma^2)T}{\sigma\sqrt{T}}$$

$$= d_1 - \sigma\sqrt{T}.$$

A proof of this is given in the CRR paper, but we give an alternative proof in Appendix A on **The Binomial Distribution**, where we use the **Berry-Esséen Theorem**.

For large N

$$\Phi[a; N, \pi'] \approx \mathcal{N}(d_1)$$
$$\Phi[a; N, \pi] \approx \mathcal{N}(d_2).$$

78   4 Multiperiod Binomial Models

Remark 4.9. The Black and Scholes formula, or perhaps better, the Black, Scholes and Merton formula, was presented in Black and Scholes [8] and Merton [53] using continuous time stochastic calculus methods. F. Black died in 1995, but M. Scholes and R. Merton received the Nobel Prize for Economics in 1997 for this work. In the 1970s, economists were not conversant with the mathematical tools used, so Cox, Ross and Rubinstein, wrote the paper [18] in which they rederived the Black and Scholes formula as a limit from the binomial model. The CRR paper is one of the most cited papers in the finance literature.

4.11 Exercises

Exercise 4.10. Show how formulae (4.7) and (4.8) follow from (4.4).

Exercises 4.11–4.14 use the data: $S(0,0) = 90$, $r = 8\%$, $T = 1$, $N = 10$, $\Delta t = 0.1$, and various choices for σ.

Exercise 4.11. Produce a spreadsheet to price the American put option with strike price $K = 100$. The answer should be \$3.08 (to the nearest cent). Now modify inputs for S, r, K = 90 and produce a modified spreadsheet. Now compute the value of this same American put option when $\sigma = 20\%, 25\%$, and find the value of σ correct to four significant figures so that the price of the American put option is \$5.00.

Calculate the European put option for the same data and various choices of σ. Also compute the early exercise premia in these cases (the difference between the American put value and the European put value).

Exercise 4.12 (The chooser option). This is also called the **As You Like It** option. The chooser option is discussed in Hull, [37] pages 461–2. Consider European call and put options both expiring at $T = 1$. A chooser option involves three dates, $t = 0$, $t = S$, and $t = T > S$. At $t = 0$ you purchase the chooser option which gives you the right to either a European put option or a European call option at $t = S$, both of which expire (with the same strike price) at $t = T$. To value the chooser option, we first compute $P(S,j)$ and $C(S,j)$, the put and call prices at $t = S$, then set $W(S,j) = \max[P(S,j), C(S,j)]$. Then compute $W(0,0)$ by usual backwardization. Use the same data as above with $K = 90$ and $S = 0.5$ (n = 5) and $\sigma = 20\%$ to evaluate the chooser option.

Remark 4.13 (The chooser option). You may be interested in a chooser option if a significant event (company takeover, etc.) were to occur at $t = S$. You may wish to have either the option to sell your stock at T for K if events are bad, or buy at T for K if events are good. Zhang [79] has the whole of Chapter 23 on the chooser option for those who wish to read more.

4.11 Exercises

Exercise 4.14 (Option on an option). Let us study the call option on a call option, although there are obviously many other combinations. These come under the title of **compound options** and were first studied by Geske [31]. We again need three dates as in (4.12). Suppose $C(T) = (S(T) - K)^+$, or in binomial tree notation $C(N, j) = (S(N, j) - K)^+$. We then compute $C(S, j)$ for each $j = 0, 1, 2, \ldots, S$, and set $W(S, j) = (C(S, j) - L)^+$. Now compute $W(0,0)$, by the usual backwardization. Evaluate this compound option with $S = 0.5$ (n = 5) and $T = 1$ (n = 10) with the data above. Let $K = 90$ and $L = 4.5$ and find $W(0,0)$. Use $\sigma = 20\%$.

Hint: Think of a way of doing both (4.12) and (4.14) with the same spreadsheet.

Remark 4.15 (Compound options). A good application of compound options is described in Hull [37] on page 443. He observes:

> The equity in a levered firm can be viewed as a call option on the value of the firm. To see this, suppose that the value of the firm is V and the face value of the outstanding debt is A. Suppose that all the debt matures at a single time, T^*. If $V < A$ at time T^*, the value of equity at this time is zero because all the company's assets go to the bondholders. If $V > A$ at T^*, the value of the equity at this time is $V - A$. Thus, the equity in the firm is a European call option on V with maturity T^* and exercise price A. An option on a stock of the firm that expires prior to T^* can be regarded as an option on an option on V.

Exercise 4.16 (The binary option). The binary call option is a European style option with payoff at expiry $W(N, j) = B > 0$ when $S(N, j) > K$ and nothing otherwise. The binary put option has payoff at expiry $W(N, j) = B > 0$ when $S(N, j) < K$ and nothing otherwise. Find the values of these options for various choices on K and B. Express the binary put option value in terms of the binary call option value.

Exercise 4.17 (Forward start option). A forward start call option is one that is paid for today and expires at $t = N$. The strike price is not known until some future time $t = M < N$. In fact the strike price $K = S(M, j)$ if state j occurs at $t = M$. With the data used in Exercises 4.11–4.14, and $M = 3$, determine the value of the forward start call option. What is the connection between the value of this option and a corresponding forward start put option?

Remark 4.18. The design of exotic options is only limited by one's imagination. Ravindran in [21, pages 81–169] devotes a large chapter to many exotic options and their applications and discusses how binomial methods can be used to estimate their values. Zhang [79] is another compendium of exotic options. As the demand is there, new financial products are designed. Such products are also **engineered** so that their present values can be estimated and (importantly) the product can be hedged (see Chapter 5).

5

Hedging

5.1 Hedging

Just as important as **pricing** is **hedging**. Some would say that hedging is even more important. Hedging is a general strategy, independent of any model. However, in this book we discuss hedging in our binomial framework.

If I sell you a European call option on asset S with strike price K and expiry T, then I should have the value $(S(T) - K)^+$ in place at time T. If you wish to exercise the option, (when $S(T) > K$), then I could buy a stock (for $S(T)$), and sell it to you for K as agreed. I would suffer a loss $S(T) - K$, but in the binomial model this is compensated for by my **hedge** which also has value $(S(T) - K)^+ = S(T) - K$ at expiry.

We now show how to hedge, or replicate, a claim. We start with an amount of cash. In each time period (see below) we divide our wealth between an investment in a bank, with interest, and the purchase of S. This portfolio can be adjusted at each time, without the addition of extra cash or the removal of any cash, but at expiry T, it must have value equal to the claim. We are then said to have hedged, or replicated, the claim. The initial cost of the hedge must be the present value of the claim. The process of hedging in this way is also called **self-financing dynamic hedging**.

We now provide some details, and give some examples. The time steps will be denoted by n, $0 \leq n \leq N$, so that N corresponds to the expiration time T.

Suppose that the general contingent claim that we are trying to hedge is denoted by W, and that $W(n, j)$ is its value at (n, j). In fact the values of $W(n, j)$ can be obtained from the values of $W(N, .)$ by the backward recursion formula (4.4).

We now write down generalizations of H_0 and H_1 which we introduced in Section 2.3, [see equation (2.2)]. We define $H_0(n, j)$ and $H_1(n, j)$ for each (n, j) so that

$$W(n+1, j+1) = H_0(n,j)R(n,j) + H_1(n,j)S(n+1, j+1) \qquad (5.1)$$

$$W(n+1, j) = H_0(n,j)R(n,j) + H_1(n,j)S(n+1, j). \qquad (5.2)$$

In fact, (5.1) generalizes (2.6), (5.2) generalizes (2.3a) and (5.2) generalizes (2.3b). We then have from (5.1) and (5.2) the formulae

$$H_1(n,j) = \frac{W(n+1,j+1) - W(n+1,j)}{S(n+1,j+1) - S(n+1,j)} \qquad (5.3)$$

$$H_0(n,j) = W(n,j) - H_1(n,j)S(n,j). \qquad (5.4)$$

Thus, at node (n,j) you can hold W (with value at (n,j) of $W(n,j)$), or equivalently $H_0(n,j)$ dollars and $H_1(n,j)$ units of S.

The quantity $H_1(n,j)$ is called the **hedge ratio**; formula (5.3) may suggest this name.

Observe that

$$W(n+1, j+1) - H_1(n,j)S(n+1, j+1) = H_0(n,j)R(n,j) \qquad (5.5)$$

$$W(n+1, j) - H_1(n,j)S(n+1, j) = H_0(n,j)R(n,j). \qquad (5.6)$$

In other words the portfolio

$$W(n+1, \cdot) - H_1(n,j)S(n+1, \cdot)$$

is **riskless**, (that is, it has the same value at both $(n+1, j+1)$ and $(n+1, j)$). Alternately, we can regard $H_1(n,j)$ as indicating **the level of exposure** of W with respect to S at (n,j).

Also note that

$$W(n,j) = H_0(n,j) + H_1(n,j)S(n,j) \qquad (5.7)$$
$$= H_0(n-1, j-1)R(n-1, j-1) + H_1(n-1, j-1)S(n,j) \qquad (5.8)$$
$$= H_0(n-1, j)R(n-1, j) + H_1(n-1, j)S(n,j). \qquad (5.9)$$

This implies

5.1 Hedging

$$[H_1(n,j) - H_1(n-1, j-1)] S(n,j)$$
$$= -[H_0(n,j) - H_0(n-1, j-1)] R(n-1, j-1),$$

so increasing our holding of stock ($H_1(n,j) > H_1(n-1, j-1)$) must be accompanied by a reduction of our cash position, and vice versa.

The conditions (5.7)–(5.9) are termed **self-financing**, that is, we reduce our cash position by purchasing shares, and we increase our cash position by selling shares.

Example 5.1 (CRR model). Suppose

$$S(0) = S = 80$$
$$u = 1.5$$
$$d = 0.5$$
$$R = 1.1$$
$$N = 3$$
$$K = 80$$

Let W be a European call option with strike price 80 expiring at $t = 3$. Thus

$$\pi = 0.6$$

and

$$W(n,j) = \frac{0.6}{1.1} W(n+1, j+1) + \frac{0.4}{1.1} W(n+1, j)$$

for $j = 1, 2, \ldots, n$.

The values of $S(n,j)$, $W(n,j)$, $H_0(n,j)$, $H_1(n,j)$ are given in Table 5.1.

The details were found using MS-EXCEL, but can be computed using other spreadsheet programmes. There are four rows in this table. They correspond to $j = 0$, $j = 1$, $j = 2$ and $j = 3$, reading down the table. The values in the final column can be entered and the earlier columns completed using (4.4), (5.3) and (5.4). We show how to use the calculations.

Suppose that the market price of the call were $36.00 (rather than $34.0796 as in the model). Here is how one can make an arbitrage profit **under the assumption that the model is correct**.

5 Hedging

Table 5.1. Hedge ratios using MS-EXCEL.

n =	0	1	2	3
S	80	40	20	10
W	34.0796	2.9752	0	0
H1	0.7190	0.1364	0	
H0	-23.4410	-2.4793	0	
S		120	60	30
W		60.4959	5.4545	0
H1		0.8485	0.1667	
H0		-41.3223	-4.5455	
S			180	90
W			107.2727	10
H1			1	
H0			-72.7273	
S				270
W				190
H1				
H0				

At t=0

Short sell the call, borrow \$23.4410 ($H_0(0,0) = -23.4410$), buy 0.7190 stock ($H_1(0,0) = 0.7190$). This realizes a profit \$1.9204, which we pocket.

We will assume the scenario $(0,0) \to (1,1) \to (2,1) \to (3,2)$.

At t=1

We are now in state $(n,j) = (1,1)$.

Repay the loan (-\$23.4410 × 1.1 = -\$25.7851), borrow \$41.3223 ($H_0(1,1) = -41.3223$), sell 0.7190 stock (+\$120 × 0.7190 = \$86.28), buy 0.8485 stock (−\$120 × 0.8485 = −\$101.82) and −25.7851 + 41.3223 + 86.28 − 101.82 = −0.0028 (actually a more accurate calculation gives 0). So these trades represent no net inflow or outflow of money. That is, our portfolio is self-financing.

In fact the **net** transactions are:

Increase the debt to \$41.3223 by borrowing \$15.5372 more.

Increase the stockholding to 0.8485, which costs 0.1295 × \$120 = \$15.54. (This would be \$15.5372 with a more accurate calculation). So again the total net transaction is \$0.

At t=2

We are now in state $(n,j) = (2,1)$.

Repay the loan (-\$41.3223 × 1.1 = -\$45.45453), borrow \$4.5455 ($H_0(2,1) = -4.5455$), sell 0.8485 stock (+\$60×0.8485 = \$50.91), buy 0.1667 stock (−\$60× 0.1667 = −\$10.002) to get −45.45453 + 4.5455 + 50.91 − 10.002 = −0.00103 (actually a more accurate calculation gives 0). So these trades represent no net inflow or out flow of money. This is again self-financing.

Actually the **net** transactions are:

Decrease debt to $4.5455 by repaying $40.90903.

Decrease stockholding to 0.1667, which gives $0.6818 \times \$60 = \$40.908$. This should be $40.90903 with a more accurate calculation. So again the total net transaction is $0.

At t=3

We are now in state $(3, 2)$.

Repay the loan (-$4.5455 $\times$ 1.1 = -$5.00005), sell 0.1667 stock (+$90 $\times$ 0.1667 = $15.003) netting $10.00295 (actually, a more accurate calculation gives $10). This is the same as the value of the call in $(3, 2)$. As the call is short held, we may now cash settle this call. We now have no unfunded liabilities at expiry.

In summary, we pocketed $1.92 at time $t = 0$, and by these trades we were able to meet all our liabilities at $t = 3$. Thus, we have an arbitrage. In a multiperiod model we need to trade periodically as we have illustrated.

We could also consider other scenarios like: $(0,0) \to (1,0) \to (2,1) \to (3,1)$.

Remark 5.2. If you **write** a call, then you collect $W(0,0)$ at time 0 and initiate the hedging procedure just described. At expiry you produce a payoff of the call and can meet the claim (against you). This is called **hedging**. All this works under the proviso that the model is correct. Inaccuracies due to a wrong model are called "model risk"!

The book by Natenburg [57] is particularly good in describing similar hedging methods to lock in arbitrage opportunities. This book is popular with practitioners, particularly with **market makers**.

Summary of Three Cases

Table 5.2. Case 1: $(0,0) \to (1,1) \to (2,1) \to (3,2)$.

Stock Position	Cash Positions (in $)
Short sell the call	36.00
Pocket	1.92
Bank	34.08
$t = 0$	34.08
$H_1(0,0) = 0.7190$	
$S(0,0) = 80$	
out $57.52	(57.52)
	(23.44)
$t = 1$	(25.78)
$H_1(1,1) = 0.8485$	
$\Delta H = -0.1295$	
$S(1,1) = 120$	
out $15.54	(15.54)
	(41.32)
$t = 2$	(45.46)
$H_1(2,1) = 0.1667$	
$\Delta H = -0.6816$	
$S(2,1) = 60$	
in $40.91	40.91
	(4.55)
$t = 3$	(5.00)
$H_1(3,2) = 0.0$	
$\Delta H = -0.1667$	
$S(3,2) = 90$	
in $15.00	15.00
	10.00

Some explanations (Table 5.2):

For case 1: At $t = 0$, $57.52 = 0.7190 \times 80$; we write "out" to indicate a cost (outflow of funds) and "in" to indicate receipts (in flow of funds) to cash position; at $t = 1$, $15.54 = 0.1295 \times 120$, $25.78 = 23.44 \times 1.1$ (as $R = 1.1$), and so on. At time $t = 3$ we have a liability of $10.00 (the value at $(3,2)$ of the call option (which we short sold), but we have now cash assets to match this. So we have no unfunded liabilities at $t = 3$. At $t = 3$ we set H_1 equal to 0, as we sell our shares at $t = 3$.

Table 5.3. Case 2: $(0,0) \to (1,0) \to (2,1) \to (3,1)$.

Stock Position	Cash Positions (in $)
Short sell the call	36.00
Pocket	1.92
Bank	34.08
$t = 0$	34.08
$H_1(0,0) = 0.7190$	
$S(0,0) = 80$	
out $57.52	(57.52)
	(23.44)
$t = 1$	(25.78)
$H_1(1,0) = 0.1364$	
$\Delta H = -0.5826$	
$S(1,0) = 40$	
in $23.30	23.30
	(2.48)
$t = 2$	(2.73)
$H_1(2,1) = 0.1667$	
$\Delta H = 0.0303$	
$S(2,1) = 60$	
out $1.82	(1.82)
	(4.55)
$t = 3$	(5.00)
$H_1(3,1) = 0.0$	
$\Delta H = -0.1667$	
$S(3,1) = 30$	
in $5.00	5.00
	(0.00)

Some explanations (Table 5.3):

For case 2: At time $t = 3$ we have no liability (the value at $(3, 2)$ of the call option (which we short sold) is $0). So we have no unfunded liabilities at $t = 3$.

Table 5.4. Case 3: $(0,0) \to (1,1) \to (2,2) \to (3,3)$.

Stock Position	Cash Positions (in $)
Short sell the call	36.00
Pocket	1.92
Bank	34.08
$t=0$	34.08
$H_1(0,0) = 0.7190$	
$S(0,0) = 80$	
out $57.52	-57.52
	-23.44
$t=1$	-25.78
$H_1(1,1) = 0.8485$	
$\Delta H = 0.1295$	
$S(1,1) = 120$	
out $15.54	-15.54
	-41.32
$t=2$	-45.46
$H_1(2,2) = 1.0$	
$\Delta H = 0.1515$	
$S(2,2) = 180$	
out $27.27	-27.27
	-72.73
$t=3$	-80.00
$H_1(3,3) = 0.0$	
$\Delta H = -1.0$	
$S(3,3) = 270$	
in $270.00	270.00
	190.00

Some explanations (Table 5.4):

For case 3: At time $t = 3$ we have a liability—the value at $(3, 2)$ of the call option (which we short sold) of $190.00. So we have no unfunded liabilities at $t = 3$.

5.2 Exercises

Exercise 5.3. Produce the spreadsheet for Example 5.1.

6

Forward and Futures Contracts

6.1 The Forward Contract

We now consider the forward contract in the multistep binomial model. A forward contract involves two dates, $t = 0$ and $t = N$, say. The forward contract is initiated at $t = 0$. A long forward contract is an agreement (obligation) to purchase an asset (a stock, say), at $t = N$ for an agreed price F. We write $F = F(0,0)$ to emphasize the fact that F is agreed at $t = 0$. At (N, j) the (long) forward contract is worth $S(N, j) - F(0,0)$. The value of this at $t = 0$ is $S(0,0) - F(0,0)P(0, N)$. This should have a value at $(0,0)$ of 0 for a fair value of $F(0,0)$. Thus,

$$F(0,0) = \frac{S(0,0)}{P(0,N)}. \tag{6.1}$$

The value V at (n, j) of the (long) forward contract initiated at time 0 with maturity N is

$$V(n,j) = S(n,j) - \frac{S(0,0)}{P(0,N)}P_j^n(N-n). \tag{6.2}$$

In general, if $F(n, j)$ is the $t = N$ forward price of S initiated at (n, j), then

$$F(n,j) = \frac{S(n,j)}{P_j^n(N-n)}. \tag{6.3}$$

Note that $V(n, j)$ is not zero in general for $0 < n < N$.

Remark 6.1. A forward contract is a binding agreement. However **the short side bears risk of default**. The person who has agreed to buy at $t = N$ may not honour the contract, not turning up at maturity. However, the long side could then be subject to litigation, credit devaluation and so on. The short side could be left holding a stock for which he was promised $60, but for which the market price has fallen to $40. We could be more dramatic. Jones has agreed to buy 50 hogs from Farmer Bill on a certain Friday, but Jones does not turn up. What is Farmer Bill to do?

Note that $F(N,j) = S(N,j)$ for all $0 \leq j \leq N$.

6.2 The Futures Contract

These contracts are traded on exchanges and, as we shall see, solve the problem of default risk discussed under forward contracts.

Futures contracts are similar to forward contracts in that they are agreements to buy/sell an asset (stock) at a future date $t = N$ for an agreed price, which we shall call $G(0,0)$.

Futures contracts are standardized by futures exchanges (e.g., CBOE), and always involve **margin accounts**.

Margin Accounts

Each exchange has its own rules.

Each of the two counterparties opens a margin account with the exchange. To some extent a margin account is like any other bank account that earns interest, but it must contain a minimum amount. Further, the **clearing house** of the exchange must have access to it, in the sense that it can add and remove amounts from it on a daily basis, as we shall explain. When the amount in the margin account gets too low, a **margin call** is put out, asking that the margin account be topped up. If the call is not answered, the contract is **closed out**. For the meantime we shall assume that no margin calls are necessary (which is equivalent to all margin calls being answered).

Let us consider the long side initiated by company XYZ. Suppose XYZ opens a margin account with initial amount $M(0,0)$. The exchange will specify the minimum amount that should be placed in this margin account. This could depend on the size of the contract.

Let $G(n,j)$ denote the futures price at (n,j) for implementation at $t = N$. Of course, $G(N,j) = S(N,j)$ for each $j = 0, 1, \ldots, N$. We now wish to determine $G(n,j)$ for each $n < N$, $j = 0, 1, \ldots, n$, and in particular we want to find $G(0,0)$.

Let $M(n,j)$ be the amount in the margin account at (n,j).

In the same way there will be a short side held by company ABC. With exchange traded futures XYZ and ABC will not necessarily be known to each other. This will not matter as we shall see. Let us denote the margin account of ABC with the letter L and the amount in it at (n,j) by $L(n,j)$.

In the multistep binomial model, the **dynamics** of the margin account M are given by

$$M(n+1, j+1) = M(n,j)R(n,j) + [G(n+1, j+1) - G(n,j)] \quad (6.4)$$

$$M(n+1, j) = M(n,j)R(n,j) + [G(n+1, j) - G(n,j)], \quad (6.5)$$

and for L, on the short side

$$L(n+1, j+1) = L(n,j)R(n,j) - [G(n+1, j+1) - G(n,j)] \quad (6.6)$$

$$L(n+1, j) = L(n,j)R(n,j) - [G(n+1, j) - G(n,j)]. \quad (6.7)$$

This means that between $t = n$ and $t = n+1$ the margin account earns interest $M(n,j) \to M(n,j)R(n,j) = M(n,j)[1 + r(n,j)]$. Further, if the futures price G rises, then the increase is added to M and removed from L, and vice versa if the futures price drops. The clearing house does the adding and removal of amounts.

This process of adjusting the margin accounts in this way is called **marking to market**. The procedure of collecting and pairing variations in margin accounts is called **resettlement**. A good reference for futures markets is Duffie [26].

Computing Futures Prices

Let $\pi(n,j)$ be the risk-neutral up probability (see Chapter 4). Then using risk-neutral pricing

$$\begin{aligned}
M(n,j) &= \frac{\pi(n,j)}{R(n,j)} M(n+1, j+1) + \frac{1 - \pi(n,j)}{R(n,j)} M(n+1, j) \\
&= \frac{\pi(n,j)}{R(n,j)} [M(n,j)R(n,j) + (G(n+1, j+1) - G(n,j))] \\
&\quad + \frac{1 - \pi(n,j)}{R(n,j)} [M(n,j)R(n,j) + (G(n+1, j) - G(n,j))] \\
&= M(n,j) + \frac{\pi(n,j)}{R(n,j)} [G(n+1, j+1) - G(n,j)] \\
&\quad + \frac{1 - \pi(n,j)}{R(n,j)} [G(n+1, j) - G(n,j)].
\end{aligned}$$

This implies that

$$G(n,j) = \pi(n,j)G(n+1,j+1) + (1-\pi(n,j))G(n+1,j) \tag{6.8}$$

with

$$G(N,j) = S(N,j), \quad 0 \le j \le N. \tag{6.9}$$

This backwardization formula allows us to calculate the futures price at all nodes (n,j).

Remark 6.2. In general futures prices are not equal to forward prices, as the following example shows.

Example 6.3 (Futures and forward prices are not equal in general). Consider a binomial model with $N=3$.

The S and R values are as follows:

$S(0,0) = 100$, $S(1,0) = 87$, $S(1,1) = 115$, $S(2,0) = 75$, $S(2,1) = 100$, $S(2,2) = 133$, $S(3,0) = 65$, $S(3,1) = 87$, $S(3,2) = 115$ and $S(3,3) = 152$.
$R(0,0) = 1.0250$, $R(1,0) = 1.0300$, $R(1,1) = 1.0200$, $R(2,0) = 1.0350$, $R(2,1) = 1.0250$ and $R(2,2) = 1.0100$.

We then calculate

$$\pi(n,j) = \frac{S(n,j)R(n,j) - S(n+1,j)}{S(n+1,j+1) - S(n+1,j)}$$

$$P(n,j) \equiv P_j^n(N-n)$$
$$= \frac{\pi(n,j)}{R(n,j)}P(n+1,j+1) + \frac{1-\pi(n,j)}{R(n,j)}P(n+1,j)$$

$$F(n,j) = \frac{S(n,j)}{P(n,j)}$$

$$F(N,j) = S(N,j)$$

$$G(n,j) = \pi(n,j)G(n+1,j+1) + (1-\pi(n,j))G(n+1,j)$$

$$G(N,j) = S(N,j).$$

The calculations show that

$$F(0,0) = \$107.36$$
$$G(0,0) = \$107.12.$$

This difference is **not** due to arithmetic errors.

6.2 The Futures Contract

Value in the Margin Account at Expiry

We now compute the values of $M(N,j)$ and $L(N,j)$.
Equations (6.4), (6.5), (6.6) and (6.7) can be rewritten:

$$M(n+1,j+1) - M(n,j) = M(n,j)r(n,j) + [G(n+1,j+1) - G(n,j)]$$

$$M(n+1,j) - M(n,j) = M(n,j)r(n,j) + [G(n+1,j) - G(n,j)]$$

and

$$L(n+1,j+1) - L(n,j) = L(n,j)r(n,j) - [G(n+1,j+1) - G(n,j)]$$

$$L(n+1,j) - L(n,j) = L(n,j)r(n,j) - [G(n+1,j) - G(n,j)]$$

Example 6.4. Let us assume that the we have the scenario $(0,0) \to (1,0) \to (2,1) \to (3,1)$. Then

$$M(1,0) - M(0,0) = M(0,0)r(0,0) + G(1,0) - G(0,0)$$
$$M(2,1) - M(1,0) = M(1,0)r(1,0) + G(2,1) - G(1,0)$$
$$M(3,1) - M(2,1) = M(2,1)r(2,1) + G(3,1) - G(2,1),$$

and now add, giving:

$M(3,1) - M(0,0)$
$\quad = M(0,0)r(0,0) + M(1,0)r(1,0) + M(2,1)r(2,1) + [G(3,1) - G(0,0)]$
$\quad = M(0,0)r(0,0) + M(1,0)r(1,0) + M(2,1)r(2,1) + [S(3,1) - G(0,0)].$

Therefore,

$$M(3,1) = M(0,0) + M(0,0)r(0,0) + M(1,0)r(1,0)$$
$$+ M(2,1)r(2,1) + [S(3,1) - G(0,0)]$$

Thus, the value $M(3,1)$ is the original amount in the margin account **plus** interest earned **plus** the futures account cash settled.

If you wish to **take delivery** of the underlying then: buy the underlying for $S(3,1)$ and use the cash $S(3,1) - G(0,0)$ from the futures to convert (reduce ?) the payment to $S(3,1) - [S(3,1) - G(0,0)] = G(0,0)$.

In a similar way for the same scenario,

$$L(3,1) = L(0,0) + L(0,0)r(0,0) + L(1,0)r(1,0) \\ + L(2,1)r(2,1) + (G(0,0) - S(3,1)).$$

Using $M(0,0) = 200$ and the data in Example 6.3, we get $M(1,0) = \$190.03$, $M(2,1) = \$206.08$, $M(3,1) = \$223.73 = 200 + 200 \times 0.025 + 190.03 \times 0.03 + 206.08 \times 0.025 + [115 - 107.12]$.

Let us note that the value of $M(3,1)$ depends on the **path** that led to (3,1). For example, if we had the scenario $(0,0) \to (1,1) \to (2,0) \to (3,1)$, then $M(3,1) = \$222.34$.

Value in Margin Account Before Expiry

As above we have

$$M(n,j) = M(0,0) + \text{Interest} + [G(n,j) - G(0,0)]$$

and

$$L(n,j) = L(0,0) + \text{Interest} - [G(n,j) - G(0,0)].$$

Analysis of a Default

Suppose that the long side defaults at (n,j).

We shall first analyse the long side. The default could have occurred because $M(n,j) \leq 0$ (or below some small positive value). Then XYZ would have received a margin call. Let us assume that XYZ ignored this call. In that case the futures contract is closed out. If $M(n,j) < 0$, then XYZ will receive a bill for the debt $-M(n,j)$ (which is probably a small amount).

Now consider the short side. At (n,j), the amount that ABC has in the margin account is $L(n,j)$. ABC can now enter a futures contract expiring still at $t = N$ with another counterparty. In fact, the clearing house will arrange this new contract. In fact, ABC may not even know that a default has occurred. As a consequence

$$\begin{aligned} L(N,k) &= L(n,j) + \text{Interest}'' - [S(N,k) - G(n,j)] \\ &= L(0,0) + \text{Interest}' - [G(n,j) - G(0,0)] \\ &\quad + \text{Interest}'' - [S(N,k) - G(n,j)] \\ &= L(0,0) + \text{Interest} - [S(N,k) - G(0,0)] \\ &= L(0,0) + \text{Interest} + G(0,0) - S(N,k) \end{aligned}$$

6.2 The Futures Contract

Here Interest' denotes interest to (n,j) and Interest'' denotes interest from (n,j) to (N,k). This is the same value in the margin account as would have been there had no default occurred.

We see that the invention of marking to market and margin accounts, has "solved" the problem of defaults in forward contracts. In fact, **marking to market** is a common feature of many financial markets as a **guarantee** against default by either party. This arrangement is regarded as one of the great financial innovations of the twentieth century.

When Are Futures and Forward Prices Are Equal?

We saw in Example 6.3 that futures prices and forward prices are not equal in general. In this example $R(n,j)$ and $r(n,j)$ were state-dependent, which means that interest rates were not deterministic.

Theorem 6.5. *If interest rates are deterministic, then $F(n,j) = G(n,j)$ for all (n,j).*

Remark 6.6. We say that interest rates are deterministic if $R(n,j)$ and $r(n,j)$ for each n do not depend on j. In that case we will write $R(n)$ rather than $R(n,j)$, and $r(n)$ in place of $r(n,j)$.

Proof. Let us note that

$$F(N,j) = G(N,j) = S(N,j) \qquad (6.10)$$

for $j = 0, 1, \ldots, N$. We only need to show that when interest rates are deterministic F and G satisfy the same backwardization equation. We noted that

$$F(n,j) = \frac{S(n,j)}{P_j^n(N-n)}$$

where $P_j^n(N-n)$ is the value at (n,j) of \$1 at time $t = N$, so

$$P_j^n(N-n) = \frac{1}{R(n)} \cdot \frac{1}{R(n+1)} \cdots \frac{1}{R(N-1)}.$$

Then

$$\begin{aligned} F(n,j) &= R(n)R(n+1)\ldots R(N-1)S(n,j) \\ &= R(n)R(n+1)\ldots R(N-1) \\ &\quad \times \left(\frac{1}{R(n)} [\pi(n,j)S(n+1,j+1) + (1-\pi(n,j))S(n+1,j)] \right) \end{aligned}$$

$$= R(n+1)\ldots R(N-1)$$
$$\times \Big(\pi(n,j)S(n+1,j+1) + (1-\pi(n,j))S(n+1,j)\Big)$$
$$= \pi(n,j)F(n+1,j+1) + (1-\pi(n,j))F(n+1,j). \qquad (6.11)$$

That is

$$F(n,j) = \pi(n,j)F(n+1,j+1) + (1-\pi(n,j))F(n+1,j), \qquad (6.12)$$

which is the same backward recursion formula as (6.8) for G. From (6.10), we conclude that $F(n,j) = G(n,j)$ for all $n < N$ and $j = 0, 1, \ldots, n$. □

6.3 Exercises

Exercise 6.7. Show that the values obtained for $F(0,0)$ and $G(0,0)$ in Example 6.3 are correct. Produce a spreadsheet calculation.

Exercise 6.8. If $N = 2$ and

$$(R(1,1) - R(1,0))(S(1,1) - S(1,0)) > 0$$

show that $G(0,0) > F(0,0)$. Can you make a generalization to $N > 2$?

Exercise 6.9. If $F(0,0) > G(0,0)$, is it possible to make an arbitrage profit by taking a long position in the futures contract and a short position in the forward contract?

7

American and Exotic Option Pricing

7.1 American Style Options

We shall now exhibit one of the flexible aspects of the binomial pricing methodology, the pricing of American style options.

Recall that an American style option is one that can be exercised at any time up to and including the expiry date. These are the kind of option most frequently traded on stock exchanges.

Because of the results in Section 2.6, we shall focus on the American put option. When a stock pays dividends, it is often optimal to exercise the American call option early. We defer discussion of this situation until we have discussed the payment of dividends.

At any time prior to expiry, an American style option can be (a) exercised; (b) sold; (c) held. This implies that if V represents the value of an American put option (with strike price K), then

$$V(n,j) \geq (K - S(n,j))^+.$$

We can compute V by a simple modification of the backwardization formula. At node (n,j) we first calculate

$$W(n,j) = \frac{\pi(n,j)V(n+1,j+1) + (1-\pi(n,j))V(n+1,j)}{R(n,j)}$$

and compare $W(n,j)$ with $(K - S(n,j))^+$. If $W(n,j) > (K - S(n,j))^+$, then we do not exercise the option as it is more profitable to hold the option than to exercise it. On the other hand, if $W(n,j) \leq (K - S(n,j))^+$, then we should exercise the option. It follows then that

7 American and Exotic Option Pricing

$$V(n,j) = \max[(K - S(n,j))^+, W(n,j)].$$

Therefore, the algorithm becomes

$$V(N,j) = (K - S(N,j))^+ \tag{7.1}$$

$$W(n,j) = \frac{\pi(n,j)V(n+1,j+1) + (1-\pi(n,j))V(n+1,j)}{R(n,j)} \tag{7.2}$$

$$V(n,j) = \max[(K - S(n,j))^+, W(n,j)] \tag{7.3}$$

$$P(0,0) = V(0,0) \tag{7.4}$$

with (7.3) being the new feature.

Example 7.1. Let us compute the American put option with data as in Example 5.1 where we calculated the European call price $34.0796.

We see that the European put is worth $14.18 and the American put is worth $18.51. The difference $4.33 = $(18.51 − 14.18) is called the **early-exercise premium**.

It can also be noted that since the American put is worth more than the European put, there must have been times and corresponding states where it was optimal to exercise the American put option early. In fact these occur at (1,0), (2,0) and (2,1).

Example 7.2 (10-step example).

We again use a CRR model, as in Section 4.10,Point 4, with $\sigma = 15\%$, $r = 10\%$, $S(0,0) = 100$, $K = 100$, $T = 1$, $N = 10$.

Then

$$\Delta t = \frac{1}{10} = 0.1$$
$$u = \exp(+\sigma\sqrt{\Delta t}) = 1.048577166$$
$$d = \exp(-\sigma\sqrt{\Delta t}) = 0.953673256$$
$$R = \exp(+r\Delta t) = 1.010050167$$
$$\pi = 0.594042026$$

The American put price is 3.0762. The reader should carry out these calculations on a spreadsheet and determine the early exercise nodes. One should note the property: if it is optimal to exercise early at (n, j) then it is also optimal to exercise early at (n, k) with $k < j$ (for $n < 10$).

The nodes (2,0), (3,0), (4,1), (5,1), (6,2), (7,2), (8,3) and (9,4) constitute the early exercise frontier.

7.2 Barrier Options

These are European style options.

There are basically two types of barrier options:

1. **Knock-out options**
    These options **cease to exist** when a barrier is struck by the underlying price. There are four examples: down-and-out calls and puts, up-and-out calls and puts.
2. **Knock-in options**
    These options **come to existence** when a barrier is struck by the underlying price. There are four examples: down-and-in calls and puts, up-and-in calls and puts.

While barrier options may once have been regarded as **exotic options**, they are now rather commonplace. They are not traded on exchanges but in the **over the counter** (OTC) market. Barrier options often involve currencies and are issued by banks who have the technology to price and hedge them. Barrier options were introduced because plain vanilla options are often too expensive and features of the vanilla options may not match client requirements. This will be apparent in the examples below. In fact most exotic options and various investment products are introduced because there is a demand for such products. You will recall our earlier discussion on the reason why there is a market for call options. There is no point introducing a **brown bear option** if there are no buyers or sellers.

Example 7.3 (Down and out call option). A Canadian company ABC is to pay 1 million USD for some imports in 3 months time, and ABC is concerned that the exchange rate will go down. We are talking here of the exchange rate that is usually quoted on the news—the indirect quote. This means that the CAD value of the import will rise. What is ABC to do? There are various solutions.

ABC could purchase a **European call option** with face value 1 million USD and a strike price $K = 1.39$ (this is the direct quote corresponding to about 72 cents). If the CAD goes down (that is, the direct quote goes up) we are guaranteed that ABC has to pay at most 1,390,000 CAD for the imports. However, **the premium for this call option may be too large**. ABC thus looks for a cheaper way to obtain a similar protection.

ABC could instead purchase a **down-and-out call option**. This is the same as above but with a knockout barrier $B = 1.35$, say. This corresponds to the market quote of 74 cents. If the CAD rose to 74 cents, then the option ceases to exist. This is the same as saying in direct terms: When the exchange rate goes down to 1.35, the option ceases to exist. After the knockout, the option no longer exists, and so there is no longer any protection against a falling dollar (in market terms). ABC will have to **take the chance** that if the CAD rose to 74 cents then it will not fall below 72 cents afterwards. It should be clear that, as the barrier option offers less protection than the plain vanilla call option, that the barrier option should be cheaper.

Let us discuss some terminology.

Customization. Barrier options are customized to the user so that they better fulfill the user's needs. For example, you wish to forgo insurance when you do not think it will be needed. Perhaps insurance companies could offer cheaper insurance with a product that offers no insurance against fires in winter nor against floods in summer.

Barrier Monitoring. This is a new feature of barrier options. The question to be answered is this: **How does one determine that the barrier has been crossed**? End-of-day monitoring would mean that the barrier is deemed to have been crossed if it is crossed at the end of a trading day. There is also continuous monitoring, but this may be difficult to implement. The type of monitoring will have an effect on the barrier option price. However, we leave this subject to the interested reader. In our binomial models, we shall monitor at nodes in the tree.

Specifications. For a barrier option we need to specify which of the eight types it is, the strike price, the barrier, the expiry date and, if needed, the face value.

Many banks offer barrier options and employ **quants** to price them. You need to be good at computer programming to do this.

Example 7.4 (Up-and-out call option). This example uses the CRR model for a stock price.

Let $S(0,0) = 100$, $K = 80$, $\sigma = 0.20$, $T = 1$, $N = 4$, $r = 0.05$; then

$$\Delta t = \frac{T}{N} = 0.25$$
$$u = \exp(\sigma\sqrt{\Delta t}) = 1.105170918...$$
$$d = \exp(-\sigma\sqrt{\Delta t}) = 0.904837418...$$
$$R = \exp(r\Delta t) = 1.012578452...$$
$$\pi = \frac{R-d}{u-d} = 0.537808372...$$

7.2 Barrier Options

The barrier will be set at $B = 120$. When the stock price reaches this level, the call option ceases to exist.

The valuation of the barrier option is the same as for the plain vanilla option except for some modifications. Let V denote the value of this option. Then

$$V(N,j) = \begin{cases} S(N,j) - K)^+ & \text{if } S(N,j) < B \\ 0 & \text{if } S(N,j) \geq B. \end{cases}$$

We now locate the barrier nodes. For each n these are the nodes (n,j) with smallest j such that $S(n,j) \geq B$. The value 0 is given to such barrier nodes. We then use the backwardization formula to find all other values of $V(n,j)$, where $S(n,j) < B$.

Thus, $V(4,j) = (S(4,j) - 80)^+$ and $V(4,3) = V(4,4) = 0$ as $S(4,3) = 122 > 120$ and $S(4,4) = 149 > 120$. The only barrier node is $(2,2)$ and we set $V(2,2) = 0$. We eventually obtain $V(0,0) = \$6.26$. If there were no barrier then the call option would cost $24.64. So you can see the savings involved.

The algorithm can be summarized as follows:

$$V(N,j) = \text{IF}\left(S(N,j) < B, (S(N,j) - K)^+, 0\right)$$

$$W(n,j) = \frac{\pi(n,j)V(n+1,j+1) + (1 - \pi(n,j))V(n+1,j)}{R(n,j)}$$

$$V(n,j) = \text{IF}\left(S(n,j) < B, W(n,j), 0\right),$$

where IF (P, A, B) means: if P is true take value A, else B.

These are formulae used (in the spreadsheet, say) to compute the barrier option prices.

Example 7.5 (Up and in call option). You may be an investor/speculator and you think that stock prices will rise from $S(0,0) = \$100$. That is, you are bullish. One possible strategy could be to buy a call option with strike $K = 95$, say. We have already talked about leverage in an earlier lecture. It is however cheaper to buy a **knock-in call option**, (an up-and-in call option). This option does not come into existence until the stock price rises to a barrier level, which we shall set at $B = 120$. It goes without saying that the premium for this barrier option will be less than that of the vanilla option. Cheaper options can provide greater leverage!

Let us price this option using the same data as in Example 7.4 above.

Again let V denote the value of this option.

We **first** find the value of V at the **barrier nodes** (the ones on or just above the barrier). At these nodes the call option is calculated as it is for the vanilla call option. Thus $V(2,2) = \$29.49$. At expiry we set $V(4,j) = 0$ if $S(4,j) < B$. We then calculate $V(n,j)$ for all nodes below the barrier in the usual way once the boundary node values have been assigned.

It is useful therefore to compute the vanilla call option price, C, as well as the knock-in call price.

If we let C be the vanilla call price, then

$$C(N,j) = (S(N,j) - K)^+$$
$$C(n,j) = \frac{\pi(n,j)C(n+1,j+1) + (1-\pi(n,j))C(n+1,j)}{R(n,j)},$$

from which V can be calculated thus:

$$V(N,j) = \text{IF}\,(S(N,j) < B,\ 0,\ C(N,j))$$

$$W(n,j) = \frac{\pi(n,j)V(n+1,j+1) + (1-\pi(n,j))V(n+1,j)}{R(n,j)}$$

$$V(n,j) = \text{IF}\,(S(n,j) < B,\ W(n,j),\ C(n,j))\,.$$

These formulae were used to compute this option. The vanilla option price was \$13.51, while the knock-in option cost \$12.03.

These two examples show the techniques that can be used to price barriers.

Some tricks can be employed. For example it is easy to see that a knock-in plus a knock-out is the same as a vanilla option. In fact if in Example 7.3 we had used $K = 80$ as in Example 7.2 then the knock-in call would cost \$18.21 and now we see that \$18.21 plus \$6.26 equals \$24.46 as we claimed. The reader may wish to check these calculations.

7.3 Examples of the Application of Barrier Options

The purpose of this section is to give you several more applications of barrier options. Read through these examples. Do not be too concerned if the various exchange rates are not the current rates. We wish to concentrate on ideas here, not give financial advice.

7.3 Examples of the Application of Barrier Options

Example 7.6 (Down-and-out call). Consider a U.S. company that must make a payment in 51 days time of 200 million Euros. The company is concerned that the Euro may appreciate against the U.S. dollar over this period. To insure against this risk, the company could purchase an ordinary foreign currency European call option, allowing it to purchase Euros at a stated price. Over the next 51 days, it is possible that the Euro depreciates against the dollar. If this happens the company may feel it no longer wants the insurance provided by an ordinary call option. What should the company consider instead? Discuss how it works.

Discussion

Suppose the current USD/Euro exchange rate is about 1.09, so 200 million Euros is about 218 million USD. You are concerned that the Euro may appreciate against the USD. If it went up to 1.11 your bill in 51 days time would increase by 4 million USD. There are various ways of dealing with future uncertainty.

1. Do nothing and pay the 200 million Euros in USD at the going exchange rate.

2. Enter a forward contract to pay the 200 million Euros in USD at the forward exchange rate. You will then know what you must pay in 51 days.

3. Purchase a European call option on 200 million Euros with strike price 1.09 (ATM), expiring in 51 days. You then pay a premium (the cost of the option) and at expiry this call is worth $200(X(51) - 1.09)^+$ million USD. Here, $X(51)$ denotes for convenience the USD/Euro exchange rate in 51 days. You therefore have the right to purchase the 200 million Euros for 1.09 USD per Euro. Thus, the most you have to pay is $200 \times 1.09 = 218$ million USD. However the premium may be too large for you to consider this possibility.

4. Purchase a European call option on 200 million Euros with strike price 1.09 (ATM), expiring in 51 days but with knock-out (down-and-out) at 1.06, say. This means that if the USD/Euro exchange rate falls to 1.06, the call option ceases to exist. This option will be cheaper than the one in 3. If you set the barrier at 1.07 it will be cheaper still. You will have to make a choice of where to set it. This will depend on the premium you are prepared to pay and the risks you are prepared to take. The gamble you take with this barrier option is this: If the exchange rate falls to 1.06, the exchange rate will not then rise above 1.09. You will have to decide (based on advice) whether this is a reasonable gamble.

Example 7.7 (Down-and-in put). Consider now a company that expects to receive a payment of 200 million Euros in 47 days time. The company is concerned that the Euro may depreciate against the dollar. One way of insurance would be for the company to purchase a put option on the Euro. Suppose the

Euro has been appreciating against the dollar and the company expects this general trend to continue. Can the company do better than the vanilla put? Discuss how it works.

Discussion

In this case you are to receive 100 million Euros in 47 days time. This will be about 218 million USD. In this case you do not mind if the Euro rises. If it rose to 1.11 you would get an extra 4 million USD. However, you are now concerned that the Euro will fall in value. If it fell to say 1.07, you would be 6 million USD worse off. We could repeat the discussion of Example 7.6, but in summary, we could now consider buying a down-and-in European put option with strike at 1.09, expiring in 47 days time, with a barrier at 1.06. This means that the put will not exist until the exchange rate actually falls to 1.06. You are supposing this is not going to happen, but you do not want to take chances in case it does. The lower you set the barrier, the cheaper the put option is going to be. So if the exchange rate falls to 1.06 you will get at least 212 million USD. If it does not reach the barrier you may get only a little more than 212 million USD.

Example 7.8 (Up-and-out put). Consider a U.S. company that is receiving a payment of 100 million Japanese yen in 30 days time. The current spot exchange rate is $0.0085 per yen. The company faces the risk that the yen will depreciate against the dollar. Suppose the yen has been appreciating against the dollar, and the company feels that if the spot exchange rate reaches the level $0.0090 per yen, its risk exposure will be negligible. The company decides to buy an up-and-out put option, the upper barrier being $0.0090 per yen. Discuss this product. What is the situation if the barrier is/is not crossed?

Discussion

The domestic market is now U.S. Let us suppose the JPY/USD exchange rate is about 0.0085 USD/JPY. So the 100 million JPY is about 8,500,000 USD. If the exchange rate goes down, the USD company will lose value. For example if the exchange rate falls to 0.0080 USD/JPY the loss would be 50,000 USD. Buying an up-and out put with barrier at 0.0090 USD/JPY means that the put option ceases to exist once the exchange rate rises to this level. If this put option has strike rate 0.0085 USD/JPY, then the U.S. company will get at least 850,000 USD provided the exchange rate did not go above 0.0090 USD/JPY, and takes a gamble that the exchange rate will not fall below 0.0085 USD/JPY if at some time it rises to 0.0090 USD/JPY. By taking this (slight) gamble, the U.S. Company is able to buy a much cheaper protection against a falling USD/JPY exchange rate.

Example 7.9 (Up-and-in call). Consider a U.S. company that must pay the principal of 50 million Euros on a Eurobond that matures in 76 days time. The company is concerned that the Euro may appreciate against the dollar. However the company has observed that the Euro has been depreciating

7.3 Examples of the Application of Barrier Options

against the dollar, and it expects this decline to continue over the near term. If the company's expectations are correct, there would be no need for the company to buy insurance against the appreciation of the Euro. Why could the company solve its problem with the purchase of an up-and-in call option?

Discussion

At the moment 1 USD is about 1.09 Euros. If this is the case then the directly quoted rate for the Euro would be about 0.92. Thus 50 million Euros is about 46 million USD. If the Euro/USD rose to say 0.95, then you would need to pay an extra 1.5 million USD. If the company expects the Euro/USD exchange rate to fall, there is no need for insurance. The company could consider buying an ATM European call (strike rate 0.97) on the 50 million Euros, expiring in 76 days, with a barrier at 0.99, say. If the Euro/USD exchange rate continues to fall, the the call option will not knock in. However, if the exchange rate were to rise above 0.99 then it knocks in and you will have to pay at most 48.5 million USD. However, if the barrier is not reached, you will pay at most 49.5 million USD (when the exchange rate rises to just below 0.99 but the barrier is never breached). That is the risk that is being taken by purchasing the cheaper option.

Example 7.10 (Up-and-out call). Consider a modification of the example of the knock-out call (up-and-out call). Suppose that the value of the up-and-out call is \$X. The additional feature to be considered here is this. If the option knocks out, you get your money back (the \$X). What should be the value of X?

Discussion

Please refer to the Example 7.4.

Let $V(n,j)$ be the value of this option in state (n,j). Without the rebate we used $V(4,0) = 0$, $V(4,1) = 1.87$, $V(4,2) = 20$, $V(4,3) = 0$, $V(2,2) = 0$, together with

$$V(n,j) = \frac{1}{1.0125}[0.5378V(n+1,j+1) + 0.4622V(n+1,j)] \quad (7.5)$$

to compute $V(0,0) = 6.26$. (In fact $V(3,2) = 9.1299$, $V(3,1) = 11.4769$, $V(3,0) = 0.9933$, $V(2,1) = 10.0885$, $V(2,0) = 6.5495$, $V(1,1) = 4.6054$, $V(1,0) = 8.3484$, $V(0,0) = 6.2572$).

So far, these were the details provided in Section 7.2.

Let $W(n,j)$ also satisfy (7.5), but $W(4,0) = W(4,1) = W(4,2) = 0$, $W(4,3) = 1$, $W(2,2) = 1$. Then with $\pi = 0.5378$, $W(3,0) = W(3,1) = 0$, $W(3,2) = \frac{\pi}{R}$. Then $W(2,0) = 0$ and $W(2,1) = \frac{\pi^2}{R^2}$. Then $W(1,0) = \frac{\pi^3}{R^3}$, $W(1,1) = \frac{\pi}{R} + \frac{(1-\pi)\pi^2}{R^3}$, and so

$$W(0,0) = \frac{\pi^2}{R^2} + \frac{(1-\pi)\pi^3}{R^4} + \frac{(1-\pi)\pi^3}{R^4}$$

$$= 0.2821 + 0.0684 + 0.0684 = 0.4189.$$

So we have two expressions for the option with rebate:

$$X = 6.2572 + 0.4189X, \tag{7.6}$$

which implies that $X = 10.7679$. We can modify the analysis. Suppose you are granted $\alpha\%$ of the premium as rebate when the option knocks out, then equation (7.6) becomes

$$X = 6.2572 + 0.4189\alpha X \tag{7.7}$$

and so

$$X = \frac{6.2572}{1 - 0.4189\alpha}. \tag{7.8}$$

$W(0,0)$ is the present value of a 1 rebate at each barrier node.

7.4 Exercises

Exercise 7.11. Verify the results in Example 7.1 and study the early exercise possibilities.

Exercise 7.12. Verify the results in Example 7.2 and study early exercise. This needs a ten-step binomial tree.

Exercise 7.13. For the CRR binomial model for S, let $V(n, j)$ be the value of the American put option at time n in state j. It expires at $t = N$ and the strike price is K. Show the following results.

(a) $V(n, j) \geq V(n + 1, j) + S(n + 1, j) - S(n, j)$.
(b) $V(n, k) \geq V(n, j) + S(n, j) - S(n, k)$ for $k > j$.
(c) If there is no early exercise at $(n + 1, j)$, then there is no early exercise at (n, j).
(d) If there is no early exercise at (n, j), then there is no early exercise at (n, k) for $k > j$.
(e) If early exercise occurs at (n, j), then it will also occur at (n, k) for $k < j$.

(f) If early exercise occurs at (n, j), then it will also occur at $(n+1, j)$.

As a hint, show that (a) and (b) hold for $n = N-1$ and $n = N$, respectively. Then use the backwardization formula for American put option valuation to obtain these inequalities for the smaller values of n.

Exercise 7.14. Verify the results in Example 7.4.

Exercise 7.15. Verify the results in Example 7.5.

Exercise 7.16. Study Section 7.3 and verify the results in Example 7.10.

Exercise 7.17 (The booster option). The booster option has two barriers, called L and H (for **L**ow and **H**igh). The booster option pays the holder an amount that is proportional to the time that the stock price (say) stays between the barriers. For simplicity, let us suppose that when the barrier H is reached at $t = n$, then the booster ceases to exist and the holder is paid \$n. The same would apply to the lower barrier L. If we let W denote the value of the booster, then $W(N, j) = N$ if $L < S(N, j) < H$, we set $W(n, j) = n$ if either $S(n, j) \geq H$ or $S(n, j) \leq L$. All other values of W are found by backwardization. Find $W(0, 0)$ for the data above for $\sigma = 15\%, 20\%, 25\%, 30\%$. Comment on your answers. Can you give an explanation, and a possible application for a booster option. Use $L = 85$ and $H = 115$

Remark 7.18 (The booster option). The booster option and its evaluation in the Black and Scholes framework is given in Chesney and Loubergé [12] and also in Dana and Jeanblanc [20].

Exercise 7.19 (Partial barrier option). Consider Example 7.4. Repeat this example, but make the assumption that knock-out can only occur at $t = 0.75$ and $t = 1$. What is the new value of this barrier option.

8

Path-Dependent Options

We now show how to modify our binomial tree methods to deal with path dependent options. We shall illustrate the ideas with **Asian options** (also called **Average Rate Options** or just **AROs** for short), and the **lookback options**. These are the most common examples.

These examples need to be analyzed using **non-recombining binomial trees**. This will mean that at time $t = n$ there will be 2^n states of the world as opposed to $n+1$ states with recombining trees. It is known that $2^n > n+1$. This causes problems as the number of states is growing exponentially. There are special techniques to help with this problem which, a good practitioner must know about, but a discussion of them is beyond the scope of this book. We shall confine our examples to ones with small n (say n is 4 or 5). Those who wish to explore this important issue further should look at J. Hull and A. White [36]. A brief outline of their technique is also in Hull's book [37] in the section on "Path-Dependent Derivatives" and in this chapter in section 8.5.

8.1 Notation for Non-Recombing Trees

We shall label the nodes of the tree as (n, j), but j will be written in base 2. (You may see now what you could do with trinomial trees).

So we have $(0,0), (1,00), (1,01), (2,000), (2,001), (2,010), (2,011)$ and so on. What is nice about this notation is that with $(2, 010)$ we can work out the past history: Start at 0, then up (1), then down (0). Also note that $0 = 000_2$, $1 = 001_2$, $2 = 010_2$, $3 = 011_2$, and so on, where the suffix 2 means a positive integer M is written in base 2 as $M = (a_p a_{p-1} \ldots a_0) = a_p 2^p + a_{p-1} 2^{p-1} + \cdots + a_0 2^0$.

With this notation, if we are at (n, j) at time $t = n$, then at time $t = n+1$ we are either at $(n+1, j1)$ (up) or at $(n+1, j0)$ (down). We remark that there are 2^n states at time $t = n$, as there are 2^n such **state labels** each having $n + 1$ digits.

With this notion all our earlier formulas do not change much. In fact

$$u(n,j) = \frac{S(n+1, j1)}{S(n,j)} \tag{8.1}$$

$$d(n,j) = \frac{S(n+1, j0)}{S(n,j)} \tag{8.2}$$

$$\pi(n,j) = \frac{R(n,j) - d(n,j)}{u(n,j) - d(n,j)} \tag{8.3}$$

$$W(n,j) = \frac{\pi(n,j)W(n+1, j1) + (1 - \pi(n,j))W(n+1, j0)}{R(n,j)} \tag{8.4}$$

where (8.4) would be suitably modified for American style options.

8.2 Asian Options

For the European call Asian option, the payoff at expiry is given by

$$W(N,j) = (A(N,j) - K)^+ \tag{8.5}$$

and for the European put Asian option, the payoff at expiry is given by

$$W(N,j) = (K - A(N,j))^+ . \tag{8.6}$$

Here $A(N,j)$ is the average value of S over the time from $t = 0$ to $t = N$. This average will depend on the path that was followed from $(0,0)$ to (N,j). Hence the payoff in (8.5) is path-dependent.

We can write

$$A(N,j) = \frac{S(0,0) + S(1, j_1) + \ldots + S(n, j_n) + \ldots + S(N, j)}{N + 1} \tag{8.7}$$

for any admissible path through the tree

$$(0,0) \to (1, j_1) \to \ldots \to (n, j_n) \to \ldots \to (N, j).$$

With the notation of the previous section, the state label j_n is obtained from the state label of (N, j) by truncation to the first $n + 1$ digits.

8.2 Asian Options

These average values can also be calculated recursively (by forward induction) as follows:

$$A(0,0) = S(0,0)$$
$$A(n,j1) = \frac{n \cdot A(n-1,j) + S(n,j1)}{n+1}$$
$$A(n,j0) = \frac{n \cdot A(n-1,j) + S(n,j0)}{n+1}.$$

Application

We now illustrate the application of the Asian option.

Suppose a Canadian company MML imports one million USD worth of widgets at the end of March, June, September and December each year. The exchange rates are $X(0.25)$, $X(0.50)$, $X(0.75)$ and $X(1.00)$, so the annual bill is

$$1000000(X(0.25) + X(0.50) + X(0.75) + X(1.00))$$

CAD. If $X(0.25) = 1/0.68$, $X(0.50) = 1/0.73$, $X(0.75) = 1/0.71$ and $X(1.00) = 1/0.72$, then the total bill is 5,637,790 CAD, which corresponds to an average exchange rate of 1.4090 cents (direct) or 0.7095 cents (indirect). MML may be happy with that. If however the (direct) exchange rate rose, then the bill may become unacceptable. Suppose that MML will accept an average exchange rate of 1.4925 (direct) (0.67 indirect) implying an annual bill of about 5,970,149 CAD, but no more; then MML could take out an European call Asian option with strike price 1.4925 and face value 4 million USD. The payoff of this option is

$$4,000,000 \times \left(\frac{X(0.25) + X(0.50) + X(0.75) + X(1.00)}{4} - 1.4925\right)^+.$$

MML is, by this amount refunded (by the Merchant Bank, say) for any annual expense over 5,970,149 CAD.

An alternative would be for MML to take out four call options expiring at $t = 0.25$, $t = 0.50$, $t = 0.75$, $t = 1.00$, each with face value one million USD and strike price $K = 1.4925$. It can be shown that the sum of the value of these four calls is greater than the cost of the Asian or average rate option. Roughly speaking this is because the average of the four rates is less volatile than each of the separate exchange rates. Another possibility would be for MML to take out one call with face value 4 million USD, strike price $K = 1.4925$ with expiry at $t = 1$. This also is more expensive than the ARO, as we shall see.

8 Path-Dependent Options

Example 8.1. A spread-sheet can be used for the valuation of an average option.

The input data were $S(0,0) = 100$, $r = 5\%$, $\sigma = 20\%$, $T = 1$, $N = 4$, and the prices for S evolve as in the CRR tree. The price of the Asian call is \$5.6661, and the price of the corresponding vanilla call is \$9.0469, which is nearly twice the price. There is also a call-put parity formula for Asian options:

$$\text{Asian Call}(0) - \text{Asian Put}(0) = S(0) - PV_0(K), \tag{8.8}$$

which is very similar to the formula for the vanilla options. The Asian Put price with the same data was \$0.7891 as compared with \$4.1698 for the vanilla put option.

Variants

Many variants can be made for the payoff of the Asian option. Here are some:

8.3 Floating Strike Options

$$W(N,j) = (A(N,j) - S(N,j))^+ \tag{8.9}$$
$$W(N,j) = (S(N,j) - A(N,j))^+. \tag{8.10}$$

We could also replace A by alternative definitions. Here are examples that have been used:

$$A(N,j) = \frac{S(k,j_k) + S(k+1,j_{k+1}) + \ldots + S(N,j)}{N - k + 1} \tag{8.11}$$

$$A(N,j) = \frac{S(N,j) + S(N-1,j_{N-1}) + \ldots + S(N-k+1,j_{N-k+1})}{k} \tag{8.12}$$

In (8.12) the usual S in the payoff of a call is replaced by the average of the last k prices. This might be suitable for use in the commodity or energy area. Most of the applications seem to be in the currency area where S is really an exchange rate rather than a stock price. There are also American style versions here. There are also many ways of taking averages (arithmetic, geometric, harmonic, etc).

8.4 Lookback Options

There are two types. The payoff of the put is

$$W(N,j) = M(N,j) - S(N,j) \qquad (8.13)$$

where

$$M(N,j) \equiv M(N,j_N) = \max\{S(n,j_n) : n = 0, 1, \ldots, N\}.$$

Note that this path-dependant payoff can never be negative. This is a very desirable option to hold, but it is expensive.

Note the forward recursion for computing the maximum process M

$$M(0,0) = S(0,0)$$
$$M(n,j1) = \max[M(n-1,j), S(n,j1)]$$
$$M(n,j0) = \max[M(n-1,j), S(n,j0)],$$

which was used for the example below.

The payoff of the call is

$$W(N,j) = S(N,j) - m(N,j), \qquad (8.14)$$

where

$$m(N,j) \equiv m(N,j_N) = \min\{S(n,j_n) : n = 0, 1, \ldots, N\}.$$

So we note that payoff can never be negative. This is also a very desirable option to hold, but it is expensive.

Note the forward recursion for computing the minimum process m

$$m(0,0) = S(0,0)$$
$$m(n,j1) = \min[m(n-1,j), S(n,j1)]$$
$$m(n,j0) = \min[m(n-1,j), S(n,j0)],$$

which is used for the example below.

Applications

Lookback options are used to obtain the best prices to buy or sell assets. If you own a stock plus the lookback put, then on the sale of the stock at $t = N$ you obtain $M(N, j)$ in state j, the best price over the life of the stock. Owning a lookback call allows you to buy a stock at $t = N$ at $m(N, j)$ in state j at time $t = N$, the best deal.

Some investment houses offer products which try to attract customers by having aspects of the lookback options.

Variants

There are many variants. Here are some other payoffs:

$$W(N, j) = (M(N, j) - K)^+ \tag{8.15}$$
$$W(N, j) = (K - m(N, j))^+, \tag{8.16}$$

where (8.15) refers to the **extrema lookback call option** and (8.16) the **extrema lookback put option**.

Example 8.2. The first example is the pricing of the the lookback put with the same inputs as Example 8.1. The price is $9.6589 (compared with $4.1698 for a put with strike 100).

Example 8.3. The second example is the pricing of the the lookback call with the same inputs as Example 8.1. The price is $13.7582 (compared with $9.0469 for a call with strike 100).

Some references dealing with Exotics include R. Jarrow (ed), [42], R. Jarrow and S. Turnbull, [43] and P. G. Zhang [79].

8.5 More on Average Rate Options

We have already discussed these options under **path-dependent options**. We also provided an application of these options and some variants. These options need to be priced with **non-recombining tree** binomial models. This may work well when there are few time steps in the tree, but are less practical when the number of time steps are large, as there will be 2^N terminal nodes in an N-step tree.

In this section we shall discuss an algorithm devised by John Hull and Alan White for obtaining approximate pricing while still using recombining trees. The technique that we will describe can be used for many path-dependent

8.5 More on Average Rate Options

options. The key reference for this method was given earlier in the first section of this chapter.

We shall focus our discussion on the example given in Section 8.2, which was the **European call Asian option** with fixed strike price.

The Hull and White Method
Step 1

At each node (n, j) we compute four numbers,

$$A^1(n,j) < A^2(n,j) < A^3(n,j) < A^4(n,j) \tag{8.17}$$

defined as follows. $A^1(n,j)$ represents the smallest value of the average of "stock" prices along any path from $(0,0)$ to (n,j). In a similar way $A^4(n,j)$ represents the largest value of the average of "stock" prices along any path from $(0,0)$ to (n,j). The other values $A^2(n,j)$ and $A^3(n,j)$ are chosen so that the four numbers in (8.17) are equally spaced. We can now write down recurrence relations for these quantities.

$$A^1(0,0) = A^2(0,0) = A^3(0,0) = A^4(0,0) = S(0,0),$$

$$A^1(n+1,0) = \frac{1}{n+2}\left[(n+1)A^1(n,0) + S(n+1,0)\right],$$

$$A^1(n+1,j+1) = \frac{1}{n+2}\left[(n+1)A^1(n,j) + S(n+1,j+1)\right]$$
$$\text{for } j = 0,\ldots,n,$$

$$A^4(n+1,n+1) = \frac{1}{n+2}\left[(n+1)A^4(n,n) + S(n+1,n+1)\right],$$

$$A^4(n+1,j) = \frac{1}{n+2}\left[(n+1)A^4(n,j) + S(n+1,j)\right]$$
$$\text{for } j = 0,\ldots,n,$$

$$A^2(n+1,j) = \frac{1}{3}\left[2 \cdot A^1(n+1,j) + A^4(n+1,j)\right]$$
$$\text{for } j = 0,1,\ldots,n+1,$$

$$A^3(n+1,j) = \frac{1}{3}\left[A^1(n+1,j) + 2 \cdot A^4(n+1,j)\right]$$
$$\text{for } j = 0,1,\ldots,n+1.$$

We illustrate this with a CRR tree.

Example 8.4. We choose $S(0,0) = 100$, $\sigma = 0.2$, $T = 0.5$, $N = 5$, $\Delta = \frac{T}{N}$, $u = \exp(\sigma\sqrt{\Delta})$, $d = 1/u$.

We note that $A^1(n,n) = A^4(n,n)$ and $A^1(n,0) = A^4(n,0)$ since there is only one path $(0,0) \to (n,n)$ and only one path $(0,0) \to (n,0)$. We could use this to simplify the algorithm a little.

Step 2

We now compute the values of

$$C^i(N,j) = \left(A^i(N,j) - K\right)^+ \quad \text{for } i = 1,2,3,4 \text{ and } j = 0,1,\ldots,N.$$

We shall select $K = 100$.

Step 3

We now begin the backwardization. Let us show how to compute $C^i(n,j)$ for $i = 1,2,3,4$.

Compute first

$$M(\uparrow) = \frac{1}{n+2}\left[(n+1)A^i(n,j) + S(n+1,j+1)\right]$$
$$\in \left[A^1(n+1,j+1), A^4(n+1,j+1)\right].$$

Look at $C^2(4,2)$, then

$$M(\uparrow) = \frac{1}{6}\left[5A^2(4,2) + S(5,3)\right]$$
$$= \frac{1}{6}\left[5 \times 98.5509 + 106.5288\right] = 99.8806 \in [97.0649, 101.4277].$$

Continuing, we express

$$M(\uparrow) = \lambda A^k(n+1,j+1) + (1-\lambda)A^{k+1}(n+1,j+1)$$

with $0 \leq \lambda \leq 1$ whenever

$$M(\uparrow) \in \left[A^k(n+1,j+1), A^{k+1}(n+1,j+1)\right].$$

For $C^2(4,2)$, $M(\uparrow) \in \left[A^1(5,3), A^2(5,3)\right]$, so we solve

$$M(\uparrow) = 99.8806 = \lambda A^1(5,3) + (1-\lambda)A^2(5,3) = \lambda 97.0649 + (1-\lambda)101.4277$$

for λ. This gives $\lambda = 0.3546$. We then set

$$C^2(4,2)(\uparrow) = \lambda C^1(5,3) + (1-\lambda)C^2(5,3)$$
$$= 0.3546 \times 0.0000 + 0.6454 \times 1.4277$$

8.5 More on Average Rate Options

$$= 0.9214.$$

In general,

$$\lambda = \frac{A^{k+1}(n+1,j+1) - M(\uparrow)}{A^{k+1}(n+1,j+1) - A^{k}(n+1,j+1)}$$

and

$$C^i(n,j)(\uparrow) = \lambda C^k(n+1,j+1) + (1-\lambda)C^{k+1}(n+1,j+1)$$
$$= C^{k+1}(n+1,j+1)$$
$$+ \lambda \left[C^k(n+1,j+1) - C^{k+1}(n+1,j+1) \right].$$

We now compute $C^i(n,j)(\downarrow)$ in a similar manner:

$$M(\downarrow) = \frac{1}{n+2} \left[(n+1)A^i(n,j) + S(n+1,j) \right]$$
$$\in \left[A^l(n+1,j), A^{l+1}(n+1,j) \right]$$

and write

$$M(\downarrow) = \lambda A^l(n+1,j) + (1-\lambda)A^{l+1}(n+1,j),$$

that is,

$$\lambda = \frac{A^{l+1}(n+1,j) - M(\downarrow)}{A^{l+1}(n+1,j) - A^l(n+1,j)}$$

and then set

$$C^i(n,j)(\downarrow) = \lambda C^l(n+1,j) + (1-\lambda)C^{l+1}(n+1,j)$$
$$= C^{l+1}(n+1,j) + \lambda \left[C^l(n+1,j) - C^{l+1}(n+1,j) \right].$$

Now

$$M(\downarrow) = \frac{1}{6} \left[5A^2(4,2) + S(5,2) \right]$$
$$= \frac{1}{6} [5 \times 98.5509 + 93.8713] = 97.7710 \in [95.2115, 99.3068]$$

$$M(\downarrow) = \lambda 95.2115 + (1-\lambda)99.3068$$
$$\lambda = 0.3750$$

$$C^2(4,2)(\downarrow) = \lambda C^2(5,2) + (1-\lambda)C^3(5,2)$$
$$= 0.3750 \times 0.0000 + 0.6250 \times 0.0000 = 0.0000$$

Continuing we then set

$$C^i(n,j) = \frac{1}{R(n,j)} \left[\pi(n,j)C^i(n,j)(\uparrow) + (1-\pi(n,j))C^i(n,j)(\downarrow)\right].$$

Then

$$\pi(n,j) \equiv \pi = 0.0.5238$$
$$R(n,j) \equiv R = 1.0050$$
$$C^2(4,2) = \frac{1}{1.0050}[0.0.5238 \times 0.9214 + 0.4762 \times 0.0000] = 0.4802.$$

This procedure is repeated to compute $C^i(n,j)$ for all choices of i, n, j.

The final answer by this method is $4.58. We would not expect this answer to be very accurate as we did not use many steps. At least we have demonstrated a recombining tree method for estimating the value of an average rate option.

Because N is not large in this calculation we could also calculate the value of the average rate put option by this method. With the same data the answer is $3.82, which also would not be an accurate answer.

As there are no analytic solutions available for average rate option prices, there are no analytic values with which to compare these answers.

8.6 Exercises

Exercise 8.5. Verify the results in Example 8.1.

Exercise 8.6. Verify the results in Examples 8.2 and 8.3.

Exercise 8.7. Implement Example 8.4.

Exercise 8.8 (Ladder options). This is a European style path-dependent option determine by a finite number of ladder levels

$$L_1 < L_2 < ... < L_k.$$

To determine its payoff $W(N,j)$ at $t = N$, we first compute the maximum process value $M(N,j)$ defined in Section 8.4. If $M(N,j) < L_1$, then

$$W(N,j) = \max\left(S(N,j) - K, 0\right),$$

but if $L_n \leq M(N,j) < L_{n+1}$, then

$$W(N,j) = \max\left(S(N,j) - K, L_n - K, 0\right).$$

The reader is referred to Ravindran [21, pages 106–108] or Street [74] for more details of this product. Using the data from Exercise 8.4 and various choices for the ladder, obtain values for the ladder option.

9

The Greeks

One of the important things that a **quant** must do besides **valuing** options and knowing how to **hedge** them is to calculate the **sensitivities** of the values with respect to inputs. These sensitivities are given Greek letters, and so are called Greeks.

We shall now discuss these in turn and indicate how they are calculated with both the Black and Scholes (BS) formula and with the binomial tree models.

9.1 The Delta (Δ) of an Option

The **Delta of an option** is the **rate of change of the option's value with respect to the underlying**. For the call in BS this is

$$\Delta_C = \frac{\partial C}{\partial S} = N(d_1) \tag{9.1}$$

and

$$\Delta_P = \frac{\partial P}{\partial S} = N(d_1) - 1 \tag{9.2}$$

Let us note that for the call the Delta lies in (0,1) while for the put the Delta lies in (-1,0).

Recall the following approximation

$$f'(x) \approx \frac{f(x+h) - f(x-h)}{2h}, \tag{9.3}$$

which we shall use often. In fact, by Taylor's theorem

$$f(x+h) \approx f(x) + hf'(x) + \frac{h^2}{2}f''(x) + \frac{h^3}{6}f'''(x)$$

$$f(x-h) \approx f(x) - hf'(x) + \frac{h^2}{2}f''(x) - \frac{h^3}{6}f'''(x).$$

Now subtract and divide by $2h$. This gives

$$\frac{f(x+h) - f(x-h)}{2h} \approx f'(x) + \frac{h^2}{6}f'''(x),$$

which establishes (9.3).

In Example 7.2 the BS Deltas are 0.7709 and -0.2291.

Consider a 10-step binomial tree with $S(0,0) = 100$, $u = \exp(\sigma\sqrt{\Delta t})$, $d = 1/u$, $\sigma = 0.15$, $R = \exp(r\Delta t)$ with $r = 0.10$ and $K = 100$. These are values about which sensitivities will be calculated. For these choices the values of the call and puts expiring at $T = 1 = 10\Delta t$ are 11.51 and 1.99, respectively.

We then compute the call and put option prices at $S = 101$ and $S = 99$, subtract answers and divide by 2, using (9.3), to give approximate values $\Delta_c = (12.367 - 10.833)/2 = 0.7670$ and $\Delta_p = (1.8512 - 2.3165)/2 = -0.23265$.

Some authors introduce two previous times. Then the present is really time $n = 2$. Write $C(0) = C(2,1)$ with $S(2,1) = 100$, and $C(0) = C(2,0)$ with $S(2,0) = 100d = 95.37$, and $C(0) = C(2,2)$ with $S(2,2) = 100u = 104.86$.

Remark 9.1. In order to show (9.1) it is useful to note the identity

$$S\mathcal{N}'(d_1) = Ke^{-rT}\mathcal{N}'(d_2) \qquad (9.4)$$

with

$$\mathcal{N}'(x) = \frac{1}{\sqrt{2\pi}}e^{-\frac{1}{2}x^2}. \qquad (9.5)$$

Let us note that the hedge ratio is

$$H_1 = \frac{C(1,1) - C(1,0)}{S(1,1) - S(1,0)} = \frac{14.579 - 7.2968}{104.86 - 95.367} = 0.7671, \qquad (9.6)$$

which is the same as the Delta. In fact, in continuous hedging the option delta gives the hedge ratio, or the exposure of the option to the underlying. The Delta is then termed a **hedge parameter**.

N. Chriss [14] pages 133–140 says that for binomial trees the Delta should be calculated as in (9.6) so that it agrees with the binomial hedge ratio. Hull [37] calculates this via

$$\Delta_C \approx \frac{C(u^2 S) - C(d^2 S)}{u^2 S - d^2 S}, \qquad (9.7)$$

which gives the value 0.7718. This is very close to the BS value. There is a corresponding formula for the Delta of the put. This is supported by an article by A. Pelsser and T. Vorst, [61]. What one does depends on whether one wants to use the Delta as a sensitivity measure or as a hedge ratio.

9.2 The Gamma (Γ) of an Option

The **Gamma is the rate of change of Delta with respect to the underlying**. So for calls

$$\Gamma_C = \frac{\partial \Delta_C}{\partial S} = \frac{\partial^2 C}{\partial S^2} = \frac{n(d_1)}{S\sigma\sqrt{T}} \qquad (9.8)$$

and for puts, the same answer

$$\Gamma_P = \frac{\partial \Delta_P}{\partial S} = \frac{\partial^2 P}{\partial S^2} = \frac{n(d_1)}{S\sigma\sqrt{T}}. \qquad (9.9)$$

Here, $n(x) \equiv \mathcal{N}'(x)$.

For the example the values are 0.0202 in both cases. For the binomial tree we can use the approximation

$$f''(x) \approx \frac{f(x-h) - 2f(x) + f(x+h)}{h^2}. \qquad (9.10)$$

In fact, by Taylor's theorem (approximation)

$$\frac{f(x-h) - 2f(x) + f(x+h)}{h^2} \approx f''(x) + \frac{h^2}{24} f^{(4)}(x),$$

from which (9.10) follows.

Applying this approximation to the binomial tree, using $h = 1$,

$$\Gamma_C \approx \frac{12.317 - 2 \times 11.507 + 10.833}{1^2} = 0.186.$$

This is not close to 0.0202 and so indicates that one must be very careful. A. Pelsser and T. Vorst [ibid.] show that a good approximation for Gamma is

$$\Gamma_C \approx \frac{2}{u^2 S - d^2 S} \times \left[\frac{C(u^2 S) - C(S)}{u^2 S - S} - \frac{C(S) - C(d^2 S)}{S - d^2 S} \right] \quad (9.11)$$

and in the example this approximation gives 0.01957, which is much closer to the BS value of 0.0202. We have written $C(S)$ to mean the call price when the present (stock) price is S, etc.

Chriss [ibid., pages 175–178], gives the formula

$$\Gamma_C \approx \frac{\frac{C(2,2)-C(2,1)}{S(2,2)-S(2,1)} - \frac{C(2,1)-C(2,0)}{S(2,1)-S(2,0)}}{S(1,1) - S(1,0)} \quad (9.12)$$

and a similar formula for puts. This gives 0.02336.

Delta can be regarded as the hedge ratio. So the Gamma gives the rate at which this hedge ratio is changing with changing values of the underlyings. It is also an important tool in hedging.

9.3 The Theta (Θ) of an Option

In the Black and Scholes case we have for calls

$$\begin{aligned}\Theta_C &= \frac{\partial C}{\partial t} = -\frac{\partial C}{\partial T} \\ &= -\frac{S(0) n(d_1) \sigma}{2\sqrt{T}} - rKe^{-rT} \mathcal{N}(d_2)\end{aligned} \quad (9.13)$$

and for puts

$$\Theta_P = -\frac{S(0) n(d_1) \sigma}{2\sqrt{T}} + rKe^{-rT} \mathcal{N}(-d_2). \quad (9.14)$$

The values in Example 7.2 are -8.8143 and 0.2341.

Chriss [ibid., pages 308–312] indicates that for the binomial tree we can use

$$\Theta_C = \frac{C(2,1) - C(0,0)}{2\Delta t}. \quad (9.15)$$

In the example these are $(9.6943 - 11.50712726)/0.2 = -9.064$ and 0.07515. It can be shown that

$$\Theta + rS(0)\Delta + \frac{1}{2}\sigma^2 S(0)^2 \Gamma - rC(0) = 0, \quad (9.16)$$

so we have another formula for Θ from this by rearrangement. Actually (9.16) is the (continuous time) Black and Scholes partial differential equation (PDE) for option pricing. This could be written more precisely. Solve the boundary value problem for $u = u(t, x)$:

$$u_t(t, x) + rxu_x(t, x) + \frac{1}{2}\sigma^2 u_{xx}(t, x) - ru(t, x) = 0$$

for $x > 0$ and $t > 0$ and with final condition $u(T, x) = (x - K)^+$. Then the (call) option price at time t is given by $C(t) = u(t, S(t))$.

9.4 The Vega (κ) of an Option

Vega is not a Greek letter. Kappa (κ) (a Greek letter) was previously used in place of vega. Anyway, **vega** is the **rate of change of the option price with respect to** σ.

For the BS model we have for the call and put option

$$\kappa_C = \kappa_P = \frac{\partial C}{\partial \sigma} = S(0)\sqrt{T}n(d_1). \tag{9.17}$$

In the example $\kappa_C = \kappa_P = 30.30143$. For the binomial model we could use the approximation

$$\frac{C(0.16) - C(0.14)}{0.02} = 29.3788.$$

9.5 The Rho (ρ) of an Option

The **rho** (ρ) is the **rate of change of the option price with respect to interest rates**.

For the BS call option price

$$\rho_C = \frac{\partial C}{\partial r} = KTe^{-rT}\mathcal{N}(d_2) \tag{9.18}$$

and for the put

$$\rho_P = \frac{\partial P}{\partial r} = -KTe^{-rT}\mathcal{N}(-d_2). \tag{9.19}$$

These give values 65.4164 and -25.0673 respectively. For the binomial model we could use the approximation

$$\rho_C \approx \frac{C(0.11) - C(0.09)}{0.02} = 66.7966$$

and

$$\rho_P \approx \frac{P(0.11) - P(0.09)}{0.02} = -23.6887,$$

which is an approximation of the same order as the BS values.

Remark 9.2. While these Greeks have been presented for European calls and puts, they can be computed for any derivative product.

9.6 Exercises

Exercise 9.3. Produce a spreadsheet to implement the results on page 121 and to evaluate the Black and Scholes Greeks.

Exercise 9.4 (The Greeks). With the data $S(0,0) = 90$, $r = 8\%$, $T = 1$, $N = 10$, $\Delta t = 0.1$ and $\sigma = 22\%$, construct the 10-step tree to compute European call and put prices. You would use the spreadsheet to compute the call and obtain the put prices using call-put parity.

Now use your spreadsheet to obtain CRR estimates for the Greeks: Delta (Δ), Gamma (Γ), rho (ρ), vega; compare them with the corresponding BS values. Compute also the BS value for Theta (Θ) and show that

$$\Theta + r \cdot S(0) \cdot \Delta + \frac{1}{2} \cdot \sigma^2 \cdot S(0)^2 \cdot \Gamma - rC(0) = 0$$

10

Dividends

In this chapter we shall discuss the modelling of a dividend paying stock. It will be good to recall some basic definitions about dividends.

When dividends are paid there are three important dates. These are $t_e < t_c < t_p$. Here $t_p - t_c$ is about two weeks, $t_c - t_e$ about seven days. These three dates relate to shares traded on a stock exchange (e.g., NYSE). On date t_p, the dividend cheques are sent out. On t_c the corporation (attached to the shares) closes the books, deciding who is entitled to dividend payment, namely shareholders on the company register at that time. For shares traded on a stock exchange there is the third date (not all shares are publicly listed), t_e called the **ex-dividend date**. This means the following: if you own the share just before t_e you are entitled to the dividend, and if you sell the shares just after t_e you will still get the dividend but the buyer will not. At ex-dividend dates there should be a share price drop equal to the dividend amount (or close to that amount). Perhaps you should look for this in the financial press (ex-dividend dates are usually provided).

If the drop did not occur, you could buy the shares just before t_e and sell them just after t_e, at about the same price. This deal would entitle you to the dividends, giving an arbitrage profit.

With this preamble, we shall simplify a little. Each time $t = k$ will be regarded as an ex-dividend date at which the dividend will be denoted by D_k. Of course, if there are no dividends paid at time $t = k$, then we can accommodate this by setting $D_k = 0$. We shall see that the dividends are sometimes state-dependent as in Section 10.1, where the dividends are a percentage of the spot price; or state-independent when the dividends may be announced ahead of time of payment. Much of the discussion below can be generalized in various ways.

Our discussion will be close to that given in Chriss [14]. There is also further discussion in Hull [[37], pages 398–401] and in Chapter 5 of Cox and Rubinstein [19].

We begin with some observations about **forward prices**.

10.1 Some Basic Results about Forwards

Lemma 10.1. *Let $f(n,j)$ be the forward price negotiated at time $t = n$ in state j to buy asset S at time $t = n+1$. Then $p(n,j)$ is the risk-neutral up-probability if and only if*

$$f(n,j) = p(n,j)S(n+1,j+1) + (1 - p(n,j))S(n+1,j). \tag{10.1}$$

Proof. We use the usual notation.

The result is necessary since if π is the risk neutral up probability:

$$0 = \frac{\pi(n,j)}{R(n,j)}[S(n+1,j+1) - f(n,j)] + \frac{1 - \pi(n,j)}{R(n,j)}[S(n+1,j) - f(n,j)].$$

For the converse, we hedge the $(n+1, j+1)$ upstate Arrow-Debreu security using cash and long forward contracts. That is, we seek $H_0(n,j)$ and $H_1(n,j)$ so that

$$1 = H_0(n,j)R(n,j) + H_1(n,j)[S(n+1,j+1) - f(n,j)] \tag{10.2}$$

$$0 = H_0(n,j)R(n,j) + H_1(n,j)[S(n+1,j) - f(n,j)]. \tag{10.3}$$

Multiply (10.2) by $S(n+1,j) - f(n,j)$ and (10.3) by $S(n+1,j+1) - f(n,j)$ and subtracting the results leads to

$$H_0(n,j) = \frac{f(n,j) - S(n+1,j)}{R(n,j)(S(n+1,j+1) - S(n+1,j))} = \frac{p(n,j)}{R(n,j)}$$

by equation (10.1). But the present value of the portfolio which hedged the Arrow-Debreu security is H_0 since the present value of a forward contract is 0. However,

$$H_0(n,j) = \frac{\pi(n,j)}{R(n,j)}$$

and the result $p(n,j) = \pi(n,j)$ follows. □

Lemma 10.2.

$$f(k,j) = S(k,j) \cdot R(k,j) - D_{k+1}. \tag{10.4}$$

Proof. This follows from (10.1) and

$$S(k,j) = \frac{\pi(k,j)}{R(k,j)}[S(k+1,j+1) + D_{k+1}] + \frac{1-\pi(k,j)}{R(k,j)}[S(k+1,j) + D_{k+1}].$$

□

10.2 Dividends as Percentage of Spot Price

This means the dividends are state-dependent so the dividend paid at (k,j) is $D(k,j) = \delta(k)S(k,j)$. Suppose $\delta(k) = \delta$ (a constant). Sometimes we shall write $1 + \delta = \exp(q\Delta t)$. We have in this case

$$f(n,j) = (1 - \delta(n+1))S(n,j)R(n,j) \equiv (1-\delta)S(n,j)R(n,j). \quad (10.5)$$

This follows as

$$0 = \frac{\pi(n,j)}{R(n,j)}[(1-\delta)u(n,j)S(n,j) - f(n,j)]$$
$$+ \frac{1-\pi(n,j)}{R(n,j)}[(1-\delta)d(n,j)S(n,j) - f(n,j)]$$

and

$$S(n,j) = \frac{\pi(n,j)}{R(n,j)}[(1-\delta)u(n,j)S(n,j) + \delta u(n,j)S(n,j)]$$
$$+ \frac{1-\pi(n,j)}{R(n,j)}[(1-\delta)d(n,j)S(n,j) + \delta d(n,j)S(n,j)].$$

Consider the CRR model with cum-dividend prices $S(n,j) = S(0,0)u^j d^{n-j}$. Here $u = \exp(\sigma\sqrt{\Delta t}) = 1/d$, $R = \exp(r\Delta t)$, $\pi = (R-d)/(u-d)$.

Then the ex-dividend prices form a recombining tree of prices. These ex-dividend prices are

$$\tilde{S}(n,j) = S(0,0)(1-\delta(1))\ldots(1-\delta(n))u^j d^{n-j},$$

(where some of the $\delta(k)$ could be 0). By (10.5), π is the risk-neutral up probability for the ex-dividend tree as well.

We suppose the ex-dividend price is modelled by $\tilde{S}$:

$$\tilde{S}(n,j) = S(0,0)(1-\delta(1))\ldots(1-\delta(n))u^j d^{n-j} \tag{10.6}$$

as given above. In order to price on this tree of prices, we need to compute $\pi(n,j)$ for each (n,j). Here we use Lemma 10.1. Let

$$\pi(n,j) = \pi = \frac{R-d}{u-d}, \tag{10.7}$$

where we have assumed a CRR model. Then we claim

$$\pi\tilde{S}(n+1,j+1) + (1-\pi)\tilde{S}(n+1,j) = (1-\delta(n+1))\tilde{S}(n,j)R(n,j). \tag{10.8}$$

From (10.5), π in (10.7) is the risk-neutral probability that we should use in the backwardization formula.

To establish (10.8), we note that the left hand side is

$$\begin{aligned} S(0,0) \quad & (1-\delta(1))\cdots(1-\delta(n+1))\left[\pi u^{j+1}d^{n+1-j-1} + (1-\pi)u^j d^{n+1-j}\right] \\ &= S(0,0)(1-\delta(1))\cdots(1-\delta(n+1))u^j d^{n-j}\left[\pi u + (1-\pi)d\right] \\ &= S(0,0)(1-\delta(1))\cdots(1-\delta(n+1))u^j d^{n-j} R \\ &\equiv \tilde{S}(n,j)(1-\delta(n+1))R(n,j), \end{aligned}$$

as required.

Example 10.3. Consider a three-step CRR binomial tree with $u = 1.5$, $d = 0.5$, $R = 1.1$, $\pi = 0.6$, $\delta(1) = 0.05$, $\delta(2) = 0$, $\delta(3) = 0.06$. The European call option for $K = 60$ is $36.00 while the American call price is $37.43. So early exercise must be optimal. Early exercise is suggested at $(2,2)$. Early exercise for American call options usually occurs just before the ex-dividend, if at all. This can be seen easily with examples.

Remark 10.4. Let $V(n,j)$ denote the value at (n,j) of an American style call option written on a dividend paying stock S. Let $\tilde{S}$ denote the ex-dividend price process of S. Then

$$\begin{aligned} V(n,j) &\geq \frac{\pi(n,j)V(n+1,j+1) + (1-\pi(n,j))V(n+1,j)}{R(n,j)} \\ &\geq \frac{\pi(n,j)[\tilde{S}(n+1,j+1) - K]^+ + (1-\pi(n,j))[\tilde{S}(n+1,j) - K]^+}{R(n,j)} \\ &\geq \frac{\left[\pi(n,j)\tilde{S}(n+1,j+1) + (1-\pi(n,j))\tilde{S}(n+1,j) - K\right]^+}{R(n,j)} \end{aligned}$$

10.2 Dividends as Percentage of Spot Price

$$= \frac{[f(n,j) - K]^+}{R(n,j)}$$

$$= \left[\tilde{S}(n,j) - \delta(n+1)\tilde{S}(n,j) - \frac{K}{R(n,j)}\right]^+$$

$$> \left[\tilde{S}(n,j) - K\right]^+$$

if $\delta(n+1) = 0$ and $R(n,j) > 1$.

This shows that early exercise can occur at $t = n$ in some state only when $\delta(n+1) > 0$.

Example 10.5. In this example we show even more clearly that it is sometimes optimal to early-exercise an American call option. Consider a one-step binomial model.

$$\tilde{S}(1,1) = Su(1-\delta)$$
$$\tilde{S}(1,0) = Sd(1-\delta)$$
$$\pi = \frac{R-d}{u-d}$$
$$C(1,1) = [Su(1-\delta) - K]^+$$
$$C(1,0) = [Sd(1-\delta) - K]^+.$$

Suppose now that $Sd(1-\delta) > K$; then

$$C_E(0,0) = \frac{\pi}{R}[Su(1-\delta) - K] + \frac{1-\pi}{R}[Sd(1-\delta) - K]$$
$$= (1-\delta)S - \frac{K}{R}.$$

So $C_E(0,0) < S - K$ provides that S is large enough. In fact if S is such that

$$\delta S > K\left[1 - \frac{1}{R}\right],$$

that is,

$$d(1-\delta)S > K,$$

this implies that the call would be exercised at $t = 0$.

10.3 Binomial Trees with Known Dollar Dividends

It is no longer so easy to form a recombining tree of ex-dividend prices. However we may proceed as follows in the CRR framework.

Set

$$\mathcal{D}_0 = \frac{D_1}{R} + \frac{D_2}{R^2} + \frac{D_3}{R^3} + \ldots$$
$$\mathcal{D}_1 = \frac{D_2}{R} + \frac{D_3}{R^2} + \frac{D_4}{R^3} + \ldots$$
$$\mathcal{D}_2 = \frac{D_3}{R} + \frac{D_4}{R^2} + \frac{D_5}{R^3} + \ldots$$

and so on. These quantities $\mathcal{D}_k$ represent the value at $t = k$ of future dividends. Define

$$\tilde{S}(n,j) = (S(0,0) - \mathcal{D}_0)\, u^j d^{n-j} + \mathcal{D}_n.$$

Then

$$\tilde{S}(0,0) = S(0,0)$$
$$\pi \tilde{S}(n+1, j+1) + (1-\pi)\tilde{S}(n+1, j) = R\tilde{S}(n,j) - D_{n+1}.$$

By Lemma 10.2, the tree of prices $\tilde{S}$ behaves like ex-dividend prices and π is the risk-neutral up probability at each node. Thus, we have a recombining tree model for "lumpy dividends".

We suppose the ex-dividend price is modelled by $\tilde{S}$:

$$\tilde{S}(n,j) = (S(0,0) - \mathcal{D}_0)\, u^j d^{n-j} + \mathcal{D}_n. \tag{10.9}$$

The motivation behind this modelling is that we consider the present value $S(0,0)$ of a stock as composed of two parts: (a) $\mathcal{D}_0$—the present value of future (known) dividends, (b) $S(0,0) - \mathcal{D}_0$—the rest, which evolves in a "random way". Again we set

$$\pi(n,j) = \pi = \frac{R-d}{u-d}, \tag{10.10}$$

where we have assumed a CRR model. We now claim with this choice of $\pi(n,j)$ that

10.3 Binomial Trees with Known Dollar Dividends

$$\pi \tilde{S}(n+1,j+1) + (1-\pi)\tilde{S}(n+1,j) = \tilde{S}(n,j)R(n,j) - D_{n+1}, \quad (10.11)$$

and so, from equation (10.4) and Lemma 10.1, the π in (10.10) is the correct choice to use in the backwardization formula.

In fact the left hand side of (10.11) is

$$\begin{aligned}
&\pi\left[(S(0,0) - \mathcal{D}_0)\,u^{j+1}d^{n+1-j-1} + \mathcal{D}_{n+1}\right] \\
&\quad + (1-\pi)\left[(S(0,0) - \mathcal{D}_0)\,u^j d^{n+1-j} + \mathcal{D}_{n+1}\right] \\
&= (S(0,0) - \mathcal{D}_0)\,u^j d^{n-j}\left[\pi u + (1-\pi)d\right] + \mathcal{D}_{n+1} \\
&= (S(0,0) - \mathcal{D}_0)\,u^j d^{n-j} R + \mathcal{D}_{n+1} \\
&= R\left[\tilde{S}(n,j) - \mathcal{D}_n\right] + \mathcal{D}_{n+1} \\
&= R\tilde{S}(n,j) + \mathcal{D}_{n+1} - R\mathcal{D}_n \\
&= R\tilde{S}(n,j) - D_{n+1},
\end{aligned}$$

as required.

Remark 10.6. For convenience we have written $S(n,j)$ for the stock price when there are no dividends and $\tilde{S}(n,j)$ for the ex-dividend stock price when there are dividends. When there are no dividends, the formulas for S and $\tilde{S}$ clearly coincide.

A reason for the model (10.9) is to ensure recombining trees for the "lumpy dividends". See also Hull [ibid., pages 402–405].

Example 10.7. We price an American call option on such a dividend-paying stock.

Suppose there are inputs $S(0,0) = 100$, $D_1 = 0$, $D_2 = 1$, $D_3 = 0$ and $D_4 = 0$. Thus, with $\Delta t = 1$, $R = \exp(0.01)$, $u = \exp(\sigma\sqrt{\delta t}) = \exp(0.15)$, $d = 1/u$.

$$\begin{aligned}
\mathcal{D}_0 &= \frac{D_2}{R^2} = 0.81873075 \\
\mathcal{D}_1 &= \frac{D_2}{R} = 0.90483742 \\
\mathcal{D}_2 &= 0 \\
\mathcal{D}_3 &= 0 \\
\mathcal{D}_4 &= 0.
\end{aligned}$$

The American call price is \$45.67. In this case notice that the European and American call prices are the same, and there is no early exercise. With an American put option, there are always situations where it is optimal to exercise early; with the American call, as here, there may not be optimal choices to exercise early.

Example 10.8. We can rework this same example, Example 10.7, with inputs $D_1 = 0$, $D_2 = 10$, $D_3 = 0$ and $D_4 = 10$. Then the American and European call prices differ with $K = 80$. They are now $32.72 and $31.76, respectively. There should be an early exercise at time $n = 3$ in states $j = 2$ and $j = 3$. It is known that an early exercise in American call options occurs just before an ex-dividend date (if at all). Further discussion of early exercise of American calls can be found in Hull [[37], page 254], Cox and Rubinstein, [[19], pages 236–245] and Jarrow and Turnbull [[43], Section 7.2]. A summary from the last reference:

> The exercise of an American call option is optimal if and only if the dividend is large enough to replace the interest lost on the strike price K and the loss of the time value of the call. If the dividend is small, and the time to maturity is large, then early exercise will be unlikely

We do not digress to prove these claims, but note that in this last modification we can choose larger dividends to force an early exercise.

Remark 10.9. Let $V(n,j)$ denote again the value at (n,j) of an American style call option written of a dividend-paying stock S. Let $\tilde{S}$ denote the ex-dividend price process of S. Then

$$V(n,j) \geq \left[\tilde{S}(n,j) - \frac{D_{n+1} + K}{R(n,j)}\right]^+$$
$$> \left[\tilde{S}(n,j) - K\right]^+$$

if $D_{n+1} = 0$ and $R(n,j) > 1$.

This shows that early exercise can occur at $t = n$ in some state only when $D_{n+1} > 0$.

10.4 Exercises

Exercise 10.10. Verify the results in Example 10.3.

Exercise 10.11. Establish the results in Remark 10.4.

Exercise 10.12. Verify the results in Example 10.7.

Exercise 10.13. Verify the results in Example 10.8.

Exercise 10.14. Establish the results in Remark 10.9.

11

Implied Volatility Trees

The references for this chapter are the articles: Derman and Kani [24], Chriss [13] and Barle and Cakici [3]. The first two papers are brought together in Chapter 9 of Chriss [14].

There are, unfortunately, problems with the Black and Scholes option pricing formula and with the basic CRR formula, which is an approximation to the Black and Scholes formula. These problems arise because market data do not support the constant volatility (σ) assumption. [See Appendix C that follows.] Therefore, the Black and Scholes model does not reproduce the prices that we see in the financial press.

Both implied volatility trees (this chapter) and implied binomial trees (in the next chapter) are binomial trees that are constructed to price options consistently so they reproduce observed market prices for the options.

In the first instance we shall focus on European put and call market prices.

There are two main applications for such constructions:

1. They can be used to compute hedge ratios and the various Greeks.
2. They can be used to estimate the cost of nonstandard/exotic options written on the same underlying, together with their hedge ratios and Greeks.

The first sections deal with the original Derman and Kani constructions and the last section with the variation by Barle and Cakici. In both approaches the implied volatility structure is assumed. We shall use the Black and Scholes implied volatility, but a CRR implied volatility could also be used. In a sense, implied volatilities are just an alternative to quoting call or put prices. We transform from one to the other with the Black and Scholes formula. In practice the market provides only a finite number of option prices, and hence a finite number of implied volatility values (for various strike prices and times to maturity). An implied volatility surface is then constructed by interpolating or extrapolating from this finite number of values. Any call and put prices can be calculated using the appropriate implied volatility from this surface. In Section 11.3 we specify an explicit surface. In what follows we shall, therefore, assume that various call and put prices with different strike prices and times to maturity are known.

11.1 The Recursive Calculation

We shall present a forward recursive construction of $S(n,j)$ starting from an $S(0,0)$.

We shall assume that we have the following input data at time $t = n - 1$:

1. $S(n-1, j)$ for $j = 0, 1, 2, \ldots, n-1$.
2. $V^{put}(n-1, j)$, which is the value at $(n-1, j)$ of a put option expiring at $t = n$ with strike price $K = S(n-1, j)$ for $j = 0, 1, 2, \ldots, n-1$.
3. $V^{call}(n-1, j)$, which is the value at $(n-1, j)$ of a call option expiring at $t = n$ with strike price $K = S(n-1, j)$ for $j = 0, 1, 2, \ldots, n-1$.
4. $R(n-1, j)$ for $j = 0, 1, 2, \ldots, n-1$. This could be calculated from $R(n-1, j) = \exp(r(n-1)\Delta t_n)$ where Δt_n is the time interval (in years) between $(n-1)$th and nth steps of the tree. Here $r(n-1)$ is an interest rate given as a % per annum. These values could also come from an interest model and could be state-dependent in general.
5. $\lambda(n-1, j)$ for $j = 0, 1, 2, \ldots, n-1$, the Arrow-Debreu prices.

In Section 11.2 we discuss how to compute these inputs. We now show how to compute the values for $S(n,j)$ for $j = 0, 1, 2, \ldots, n$. There are three cases.

Case 1

Suppose we are at node $(n-1, j)$ and we know $S(n, j+1)$ and we seek $S(n, j)$. Put for convenience $K = S(n-1, j)$. Then

$$V^{put}(n-1, j) = \frac{1}{R(n-1, j)}(1 - \pi(n-1, j))\left[K - S(n, j)\right]. \quad (11.1)$$

Now we use

$$1 - \pi(n-1, j) = \frac{S(n, j+1) - R(n-1, j)K}{S(n, j+1) - S(n, j)}. \quad (11.2)$$

We can substitute (11.2) into (11.1) and rearrange the equation to solve for $S(n, j)$. This is

11.1 The Recursive Calculation

$$S(n,j) = \frac{V^{put}(n-1,j)S(n,j+1) + K\left(K - \frac{S(n,j+1)}{R(n-1,j)}\right)}{V^{put}(n-1,j) + K - \frac{S(n,j+1)}{R(n-1,j)}}, \quad (11.3)$$

or, recalling $K = S(n-1,j)$,

$$S(n,j) = \frac{V^{put}(n-1,j)S(n,j+1) + S(n-1,j)\left(S(n-1,j) - \frac{S(n,j+1)}{R(n-1,j)}\right)}{V^{put}(n-1,j) + S(n-1,j) - \frac{S(n,j+1)}{R(n-1,j)}}. \quad (11.4)$$

Case 2

Now suppose that we are at node $(n-1,j)$ and we know $S(n,j)$ and we seek $S(n,j+1)$. Let use set for convenience $K = S(n-1,j)$. Then

$$V^{call}(n-1,j) = \frac{1}{R(n-1,j)}\pi(n-1,j)\left[S(n,j+1) - K\right]. \quad (11.5)$$

Now we use

$$\pi(n-1,j) = \frac{R(n-1,j)K - S(n,j)}{S(n,j+1) - S(n,j)}. \quad (11.6)$$

We can substitute (11.6) into (11.5) and rearrange the equation to solve for $S(n,j+1)$. This is

$$S(n,j+1) = \frac{V^{call}(n-1,j)S(n,j) + K\left(\frac{S(n,j)}{R(n-1,j)} - K\right)}{V^{call}(n-1,j) + \frac{S(n,j)}{R(n-1,j)} - K}, \quad (11.7)$$

or, recalling $K = S(n-1,j)$,

$$S(n,j+1) = \frac{V^{call}(n-1,j)S(n,j) + S(n-1,j)\left(\frac{S(n,j)}{R(n-1,j)} - S(n-1,j)\right)}{V^{call}(n-1,j) + \frac{S(n,j)}{R(n-1,j)} - S(n-1,j)}. \quad (11.8)$$

Case 3

Suppose that we are at node $(n-1,j)$ and we know neither $S(n,j)$ nor $S(n,j+1)$. Let use set for convenience $K = S(n-1,j)$ and $R = R(n-1,j)$. There are various approaches here. We shall choose the **E. Derman and I. Kani method**. This is presented in [24].

Let $S(n,j+1) = Ku$ and $S(n,j) = Kd$ where $ud = 1$. Then

138   11 Implied Volatility Trees

$$V^{put}(n-1,j) = \frac{1}{R}\left(\frac{S(n,j+1)-KR}{S(n,j+1)-S(n,j)}\right)[K-S(n,j)]$$
$$= \frac{1}{R}\left(\frac{u-R}{u-d}\right)[1-d]K$$
$$= \frac{1}{R}\left(\frac{u-R}{u-\frac{1}{u}}\right)\left[1-\frac{1}{u}\right]K$$
$$= \frac{1}{R}\left(\frac{u-R}{u^2-1}\right)[u-1]K$$
$$= \frac{1}{R}\left(\frac{u-R}{u+1}\right)K,$$

which can be rearranged to

$$u = \frac{K+V^{put}(n-1,j)}{\frac{K}{R}-V^{put}(n-1,j)}.$$

Therefore,

$$u = \frac{S(n-1,j)+V^{put}(n-1,j)}{\frac{S(n-1,j)}{R(n-1,j)}-V^{put}(n-1,j)}. \tag{11.9}$$

Write

$$S(n,j+1) = S(n-1,j)\cdot u \tag{11.10}$$
$$S(n,j) = S(n-1,j)\cdot \frac{1}{u}. \tag{11.11}$$

We see which case applies in any situation.

11.2 The Inputs V^{put} and V^{call}

The inputs $P_E^{(n,K)} = P_E^{(n,K)}(0,0)$ and $C_E^{(n,K)} = C_E^{(n,K)}(0,0)$ will stand for the present ($n=0$) value of puts and calls expiring at $t=n$ with strike price K. We shall assume for the moment that we have these values.

We calculate the value for $V^{put}(n-1,j)$.

We set $K = S(n-1,j)$. For $k < j$ and use Lemmas 10.1 and 10.2 from Chapter 10:

11.2 The Inputs V^{put} and V^{call}

$$P_E^{(n,K)}(n-1,k) = \frac{1}{R(n-1,k)}[\pi(n-1,k)(K-S(n,k+1))$$
$$+(1-\pi(n-1,k))(K-S(n,k))]$$
$$= \frac{1}{R(n-1,k)}[K - S(n-1,k)R(n-1,k) + D_n],$$

where $D_n = 0$ if $t = n$ is **not** an ex-dividend date. Let us note that the right hand side of

$$P_E^{(n,K)}(n-1,k) = \frac{1}{R(n-1,k)}[K - S(n-1,k)R(n-1,k) + D_n] \quad (11.12)$$

involves known quantities. (This is even the case if we have a stochastic interest rate model.) We also have

$$P_E^{(n,K)}(n-1,k) = 0 \text{ if } k > j \quad (11.13)$$
$$P_E^{(n,K)}(n-1,j) = V^{put}(n-1,j). \quad (11.14)$$

Now bring in the Arrow-Debreu prices:

$$P_E^{(n,K)} = \sum_{k=0}^{j} \lambda(n-1,k) P_E^{(n,K)}(n-1,k)$$
$$= \sum_{k=0}^{j-1} \lambda(n-1,k) P_E^{(n,K)}(n-1,k) + \lambda(n-1,j) V^{put}(n-1,j)$$
$$= \Sigma_P(n-1,j) + \lambda(n-1,j) V^{put}(n-1,j).$$

Here

$$\Sigma_P(n-1,j) = \sum_{k=0}^{j-1} \lambda(n-1,k) \left[\frac{1}{R(n-1,k)} [K - S(n-1,k)R(n-1,k) + D_n] \right], \quad (11.15)$$

so

$$V^{put}(n-1,j) = \frac{P_E^{(n,S(n-1,j))} - \Sigma_P(n-1,j)}{\lambda(n-1,j)}. \quad (11.16)$$

The calculation of the value for $V^{call}(n-1, j)$.

This proceeds in an analogous way. Set $K = S(n-1, j)$. For $k > j$ and using Lemmas 10.1 and 10.2 from Chapter 10,

$$C_E^{(n,K)}(n-1, k) = \frac{1}{R(n-1, k)} [\pi(n-1, k)(S(n, k+1) - K) \\ + (1 - \pi(n-1, k))(S(n, k) - K)] \\ = \frac{1}{R(n-1, k)} [S(n-1, k)R(n-1, k) - D_n - K],$$

or

$$C_E^{(n,K)}(n-1, k) = \frac{1}{R(n-1, k)} [S(n-1, k)R(n-1, k) - D_n - K] \quad (11.17)$$

$$C_E^{(n,K)}(n-1, k) = 0 \quad \text{if } k < j \quad (11.18)$$

$$P_E^{(n,K)}(n-1, j) = V^{put}(n-1, j). \quad (11.19)$$

We now use the Arrow-Debreu securities so that:

$$C_E^{(n,K)} = \sum_{k=j}^{n-1} \lambda(n-1, k) C_E^{(n,K)}(n-1, k) \\ = \sum_{k=j+1}^{n-1} \lambda(n-1, k) C_E^{(n,K)}(n-1, k) + \lambda(n-1, j) V^{call}(n-1, j) \\ = \Sigma_C(n-1, j) + \lambda(n-1, j) V^{call}(n-1, j).$$

Then

$$V^{call}(n-1, j) = \frac{C_E^{(n, S(n-1,j))} - \Sigma_C(n-1, j)}{\lambda(n-1, j)} \quad (11.20)$$

where

$$\Sigma_C(n-1, j) = \sum_{k=j+1}^{n-1} \lambda(n-1, k) \left[\frac{1}{R(n-1, k)} [S(n-1, k)R(n-1, k) - D_n - K] \right]. \quad (11.21)$$

11.3 A Simple Smile Example

This is Example 1 from Chriss [ibid., page 391]. We talk about implied volatility smiles because the function expressing implied volatility in terms of the strike price is U-shaped, like a smile.

Suppose there are inputs

$$S(0,0) = 100$$
$$\Delta t_n = 1 \text{ year for all } n$$
$$r(n) = 0.05 \text{ for all } n$$
$$R(n,j) = \exp(0.05) = 1.051271096 \text{ for all } n, j$$
$$\sigma_{imp} = \begin{cases} -\frac{K}{20} + 20 & \text{if } K \leq 110 \\ -\frac{K}{10} + 25.5 & \text{if } K \geq 110. \end{cases}$$

This last statement means that if we want $P_E^{(n,K)}$ we use the Black and Scholes formula with $T = n$, $\sigma = \sigma_{imp}$, $r = 0.05$, and a similar statement for calls. In Chriss's example, the CRR option pricing formula is used, rather than Black and Scholes. For simplicity we shall use the Black and Scholes formula throughout.

Thus,

$$P_E^{(1,100)} = 4.8649 \text{ with } \sigma = 0.15$$

$$\lambda(0,0) = 1$$

$$V^{put}(0,0) = \frac{P_E^{(1,100)}}{\lambda(0,0)} = 4.8649$$

$$u = \frac{S(0,0) + V^{put}(0,0)}{\frac{S(0,0)}{R(0,0)} - V^{put}(0,0)}$$
$$= \frac{100 + 4.8649}{\frac{100}{1.05127} - 4.8649} = 1.161834$$

$$S(1,1) = S(0,0)u = 116.1834$$

$$S(1,0) = \frac{S(0,0)}{u} = 86.0708$$

$$\pi(0.0) = \frac{R(0,0)S(0,0) - S(1,0)}{S(1,1) - S(1,0)} = 0.6328$$

$$\lambda(1,1) = \frac{\pi(0,0)}{R(0,0)} = 0.6020$$

$$\lambda(1.0) = \frac{1 - \pi(0,0)}{R(0,0)} = 0.3493.$$

We are now ready to move to time t = 2.

We have already found $\lambda(1,j)$, $j = 0, 1$.

We need $P_E^{2,S(1,0)}$ and $C_E^{2,S(1,1)}$. These are computed using the Black and Scholes formula with $r = 0.05$, $T = 2$ and $\sigma = 15.70\%$ when $K = S(1,0) = 88.1344$, $\sigma = 13.88\%$ when $K = S(1,1) = 113.4629$. This gives

$$P_E^{2,S(1,0)} = 1.5860149$$

$$P_E^{2,S(1,1)} = 9.32615591$$

$$C_E^{2,S(1,0)} = 21.838626$$

$$C_E^{2,S(1,1)} = 6.6606311.$$

We set $S(2,1) = 100$. In fact we shall set $S(4,2) = S(6,3) = \cdots = 100$. Of course, this is not the only choice. Suppose [see (11.15) and (11.21)]

$$\Sigma_P(1,0) = 0$$
$$\Sigma_P(1,1) = \frac{\lambda(1,0)}{R(1,0)}[S(1,1) - S(1,0)R(1,0)] = 6.1969$$
$$\Sigma_C(1,1) = 0$$
$$\Sigma_C(1,0) = \frac{\lambda(1,1)}{R(1,1)}[S(1,1)R(1,1) - S(1,0)] = 16.8406.$$

Then

$$V^{put}(1,0) = \frac{P_E^{2,S(1,0)} - \Sigma_P(1,0)}{\lambda(1,0)} = 5.0661$$

$$V^{put}(1,1) = \frac{P_E^{2,S(1,1)} - \Sigma_P(1,1)}{\lambda(1,1)} = 4.9034$$

11.3 A Simple Smile Example

$$V^{call}(1,0) = \frac{C_E^{2,S(1,0)} - \Sigma_C(1,0)}{\lambda(1,0)} = 15.9650$$

$$V^{call}(1,1) = \frac{C_E^{2,S(1,1)} - \Sigma_C(1,1)}{\lambda(1,1)} = 10.4370.$$

Actually, not all these calculations are needed.

So, by (11.4) and (11.8),

$$S(2,0) = \frac{V^{put}(1,0)S(2,1) + S(1,0)\left(S(1,0) - \frac{S(2,1)}{R(1,0)}\right)}{V^{put}(1,0) + S(1,0) - \frac{S(2,1)}{R(1,0)}} = 56.8625$$

and

$$S(2,2) = \frac{V^{call}(1,1)S(2,1) + S(1,1)\left(\frac{S(2,1)}{R(1,1)} - S(1,1)\right)}{V^{call}(1,1) + \frac{S(2,1)}{R(1,1)} - S(1,1)} = 131.2428.$$

We can then calculate

$$\pi(1,0) = \frac{R(1,0)S(1,0) - S(2,0)}{S(2,1) - S(2,0)} = 0.8296$$

$$\pi(1,1) = \frac{R(1,1)S(1,1) - S(2,1)}{S(2,2) - S(2,1)} = 0.6171$$

$$\lambda(2,0) = \frac{1 - \pi(1,0)}{R(1,0)}\lambda(1,0) = 0.0507$$

$$\lambda(2,1) = \frac{1 - \pi(1,1)}{R(1,1)}\lambda(1,1) + \frac{\pi(1,0)}{R(1,0)}\lambda(1,0) = 0.4794$$

$$\lambda(2,2) = \frac{\pi(1,1)}{R(1,1)}\lambda(1,1) = 0.3746.$$

We have used here the forward recurrence relations of Jamshidian to compute the Arrow-Debreu prices.

We may now proceed to time t = 3.

This proceeds in the same way except for the computation of the **central** $S(3,1)$ and $S(3,2)$. These are computed by (11.9)–(11.11).

11 Implied Volatility Trees

$$u = \frac{S(2,1) + V^{put}(2,1)}{\frac{S(2,1)}{R(2,1)} - V^{put}(2,1)}$$

$$= \frac{100 + V^{put}(2,1)}{\frac{100}{R(2,1)} - V^{put}(2,1)}$$

$$S(3,2) = S(2,1) \cdot u = 100u$$

$$S(3,1) = S(2,1) \cdot \frac{1}{u} = \frac{100}{u}.$$

For this we need $V^{put}(2,1)$:

$$V^{put}(2,1) = \frac{P_E^{3,100} - \Sigma_P(2,1)}{\lambda(2,1)}$$

$$P_E^{3,100} = 4.19851762 \quad \text{using } \sigma = 0.15$$

$$\Sigma_P(2,1) = \frac{\lambda(2,0)}{R(2,0)} [S(2,1) - S(2,0)R(2,0)]$$

$$= \frac{\lambda(2,0)}{R(2,0)} [100 - S(2,0)R(2,0)] = 1.9398$$

$$V^{put}(2,1) = 4.7106$$

$$u = 1.151269614$$

$$S(3,2) = 115.1269$$

$$S(3,1) = 86.8606,$$

and now we proceed as before. Further steps are left to the reader.

11.4 In General

This procedure will work whenever we know $\sigma_{imp}(n, K)$, which will allow us to calculate $P_E^{n,K}$ and $C_E^{n,K}$ for various choices of (n, K) (given r, $S(0,0)$).

These could be calculated by the CRR formula or from the Black and Scholes formula. It depends on the **meaning** of σ_{imp} that is being used. In the example above, $\sigma_{imp}(n,K)$ did not depend on n. Sometimes we may wish to allow dependence on n as options become volatile near expiry date.

In practice $\sigma_{imp}(n,K)$ is only known for a discrete number of choices (n,K). In that case we can construct an **implied volatility surface** $\sigma_{imp}(n,K)$ by extrapolation and interpolation. Chriss [[14], pages 405–408] has a short section on **bilinear interpolation of implied volatilities**.

Remark 11.1. Sometimes we may meet problems. At each construction we need to test that the formulae give

$$S(n,j) < S(n-1,j)R(n-1,j) < S(n,j+1). \qquad (11.22)$$

There is nothing in the procedure that we have described that guarantees this. We would expect (11.22) always holds if there are no arbitrage opportunities in the market prices. Sometimes there may be (small) arbitrage opportunities that cannot be exploited, and this may imply that (11.22) is violated. If this is the case, then Derman and Kani suggest the choices:

1. If $S(n,j+1)$ is known and $S(n,j) > S(n-1,j)R(n-1,j)$, replace $S(n,j)$ from the procedure by

$$S'(n,j) = \frac{S(n,j+1)S(n-1,j-1)}{S(n-1,j)}. \qquad (11.23)$$

2. If $S(n,j)$ is known and $S(n,j+1) < S(n-1,j)R(n-1,j)$, replace $S(n,j+1)$ from the procedure by

$$S'(n,j+1) = \frac{S(n,j)S(n-1,j+1)}{S(n-1,j)}. \qquad (11.24)$$

Of course 1. arises when using **Case 1** and 2. arises when using **Case 2**.

The rationale for this modification is given in Derman and Kani [24, page 9] and in Chriss [14, pages 379–384]. We hope you will not meet these problems. When we use these modifications we must also not use the offending market prices $P_E^{(n,S(n-1,j))}$ and $C_E^{(n,S(n-1,j))}$, respectively.

11.5 The Barle and Cakici Approach

Barle and Cakici [3] claimed that the Derman and Kani approach which we have presented can generate negative risk-neutral probabilities as we have described in Section 11.4. In fact, such negative probabilities will arise whenever

$$S(n,i) > f(n,i) \tag{11.25}$$

or

$$f(n,i) > S(n,i+1) \tag{11.26}$$

for some node (n,i) [see Section 10.1]. We will describe the Barle and Cakici approach, but we will use our notation.

For (constant) interest rate r and dividend rate q we can set

$$f(n,i) = S(n,i)\exp[(r-q)\Delta t] \tag{11.27}$$
$$= \pi(n,i)S(n+1,i+1) + (1-\pi(n,i))S(n+1,i) \tag{11.28}$$

for one period forward prices [see Section 10.1]. For simplicity of presentation we set $q=0$ and write $R = \exp[r\Delta t]$. We also use the notation

$$C(K,n) = \sum_{i=0}^{n} \lambda(n,i)\,[S(n,i)-K]^{+} \tag{11.29}$$

for the price of the call option that expires at $t=n$ and the strike price is K. $P(K,n)$ for puts is defined in a similar way. Clearly $C(K,n) \equiv C_E^{n,K}(0,0)$ of previous sections. We again construct $\{S(n,j), \pi(n,j)\}$ by forward induction. We shall assume that that we have the following input data at $t = n-1$:

1. $S(n-1,j)$ for $j = 0,1,2,\ldots,n-1$.
2. $C(f(n-1,i),n)$ for $i = 0,1,2,\ldots,n-1$.
3. $R(n-1,j) = R$ for $j = 0,1,2,\ldots,n-1$.
4. $\lambda(n-1,j)$ for $j = 0,1,2,\ldots,n-1$.

Remark 11.2. We let the reader explore the use of a more general term structure for the interest rates as we did in earlier sections.

We seek now $S(n,j)$ with

$$f(n-1,j) < S(n,j+1) < f(n-1,j+1) \tag{11.30}$$

for $j = 0,1,2,\ldots,n-1$, and then

$$\pi(n-1,j) = \frac{f(n-1,j) - S(n,j)}{S(n,j+1) - S(n,j)} \tag{11.31}$$

satisfies $0 < \pi(n-1,j) < 1$ for $j = 0,1,2,\ldots,n-1$.

Let us note that

11.5 The Barle and Cakici Approach

$$C(f(n-1,i),n) = \sum_{j=i+1}^{n} \lambda(n,j)\left[S(n,j) - f(n-1,i)\right]$$

$$= \frac{1}{R} \sum_{j=i+1}^{n-1} \lambda(n-1,j)\left[f(n-1,j) - f(n-1,i)\right]$$

$$+ \frac{1}{R}\lambda(n-1,i)\pi(n-1,i)\left[S(n,i+1) - f(n-1,i)\right].$$
(11.32)

Here and below we do not give as much detail as in earlier sections. We challenge the reader to fill in the details.

Define

$$\Delta^c(n-1,i) = RC(f(n-1,i),n) - \sum_{j=i+1}^{n-1} \lambda(n-1,j)\left[f(n-1,j) - f(n-1,i)\right]$$
(11.33)

which is a known value at $t = n-1$.

Then by (11.31)

$$\Delta^c(n-1,i) = \lambda(n-1,i)\pi(n-1,i)\left[S(n,i+1) - f(n-1,i)\right]$$

$$= \lambda(n-1,i)\left[\frac{f(n-1,i) - S(n,i)}{S(n,i+1) - S(n,i)}\right]\left[S(n,i+1) - f(n-1,i)\right],$$
(11.34)

which can be rearranged to express $S(n, i+1)$ in terms of $S(n, i)$:

$$S(n,i+1) = \frac{\Delta_i^c S_i - \lambda_i[f_i - S_i]f_i}{\Delta_i^c - \lambda_i[f_i - S_i]}$$
(11.35)

where

$$\Delta_i^c \equiv \Delta^c(n-1,i)$$
$$S_i \equiv S(n,i)$$
$$f_i \equiv f(n-1,i)$$
$$\lambda_i \equiv \lambda(n-1,i).$$

In a similar way we can use equation (11.34) to express $S(n, i)$ in terms of $S(n, i+1)$:

$$S(n,i) = \frac{\lambda_i[S_{i+1} - f_i]f_i - \Delta_i^c S_{i+1}}{\lambda_i[S_{i+1} - f_i] - \Delta_i^c} \qquad (11.36)$$

with

$$S_{i+1} \equiv S(n, i+1).$$

Starting with the value of S at a central node at $t = n$ we can calculate the upper and lower values of S on the tree at $t = n$ by using equations (11.35) and (11.36).

If n is even we set

$$S(n, \frac{n}{2}) = S(0,0)R^n. \qquad (11.37)$$

but if n is odd, we shall require that

$$S(n,i)S(n,i+1) = f(n-1,i)^2 \qquad (11.38)$$

with $i = \frac{(n-1)}{2}$. This leads to

$$S(n,i) = f(n-1,i)\left[\frac{\lambda(n-1,i)f(n-1,i) - \Delta^c(n-1,i)}{\lambda(n-1,i)f(n-1,i) + \Delta^c(n-1,i)}\right]. \qquad (11.39)$$

Let us note that $f(n-1,i) < f(n-1,i+1)$ for each $i = 0, 1, 2, ..., n-2$, since we can assume $S(n-1,i) < S(n-1,i+1)$ for all these i, and equation (11.28) holds.

If the calculations do *not* yield

$$f(n-1,i) < S(n,i+1) < f(n-1,i+1),$$

then instead set

$$S(n,i+1) = \frac{1}{2}\left[f(n-1,i) + f(n-1,i+1)\right],$$

if $S(n,n) \leq f(n-1,n-1)$, set $S(n,n) = 2f(n-1,n-1) - S(n,n-1)$ and if $S(n,0) \geq f(n-1,0)$, set $S(n,0) = 2f(n-1,0) - S(n,1)$.

Once $S(n,j)$ have been constructed so that (11.30) holds, we can find $\pi(n-1,j)$ from (11.31) and hence, $\lambda(n,j)$, from the forward induction formula (4.10); $f(n,j)$ is computed from (11.27); and the $C(f(n,i), n+1)$ are calculated as described in section 11.4 from a given implied volatility surface.

11.6 Exercises

Exercise 11.3. Implement the example in Section 11.3.

Exercise 11.4 (Implied volatility). Produce a spreadsheet that will provide Black and Scholes prices for European call and put Options. Your spreadsheet should have clearly labelled cells for inputs $S(0)$, K, r, T and σ, as well as out puts d_1, d_2, $\mathcal{N}(d_1)$, $\mathcal{N}(d_2)$ together with $C(0)$ and $P(0)$ the European call and put prices. Let us now consider the following data from the AFR of Monday, July 29, 2002 for September Telstra call and put options. The spot prices was \$4.75. What is the expiry date of the September options? Calculate in years, the time from July 29 to the expiry of the September options. Use one year equals 365 days and $r = 4.92\%$ per annum as the interest rate for about 60 days (this was the 60-day BBSW rate supplied by ANZ Bank). The call option data is shown in Tables 11.1 and 11.2.

Table 11.1. Call price and implied volatility data for Exercise 11.4.

K	Call price-ask	Implied volatility-ask
4.00	0.78	na
4.25	0.54	na
4.50	0.33	14.59
4.75	0.16	16.01
5.00	0.06	16.78
5.25	0.02	18.07
5.50	0.01	22.15
5.75	0.01	22.71

Table 11.2. Put price and implied volatility data for Exercise 11.4.

K	Put price-ask	Implied volatility-ask
4.00	0.02	32.07
4.25	0.04	28.69
4.50	0.09	28.47
4.75	0.2	27.79
5.00	0.38	32.99
5.25	0.59	37.81
5.50	0.83	na
5.75	1.07	na
6.00	1.32	na
6.25	1.57	na

Now do the following:

1. Calculate the **early-exercise premium** for each of the above put options. What do you need to assume?

2. For each call and put, compute the **Black and Scholes Implied Volatility** for the options, and compare your answers with those quoted.

3. Plot σ versus K for the calls and puts. Suppose that you wish to estimate the value of a call with $K = 4.60$. What value of σ would you use in the Black and Scholes formula to compute the call price? What is your estimate for this call price?

Remark 11.5 (Interpolation method). Here is a useful interpolation method. Let $x_1 < x_2 < x_3 < x_4$ and let y_1, y_2, y_3, y_4 be arbitrary. Then the cubic polynomial that passes through $(x_1, y_1), (x_2, y_2), (x_3, y_3)$ and (x_4, y_4) is given by

$$f(x) = \frac{(x-x_2)(x-x_3)(x-x_4)}{(x_1-x_2)(x_1-x_3)(x_1-x_4)}y_1 + \frac{(x-x_1)(x-x_3)(x-x_4)}{(x_2-x_1)(x_2-x_3)(x_2-x_4)}y_2$$
$$+ \frac{(x-x_1)(x-x_2)(x-x_4)}{(x_3-x_1)(x_3-x_2)(x_3-x_4)}y_3 + \frac{(x-x_1)(x-x_2)(x-x_3)}{(x_4-x_1)(x_4-x_2)(x_4-x_3)}y_4$$

and we use this formula to compute the interpolation for $x_2 < x < x_3$.

Exercise 11.6. Construct a four-step implied volatility tree ($n = 0, 1, 2, 3, 4$) for the inputs $S(0,0) = 90$, $r = 5\%$, $\Delta t = 1$, and we have a (time-independent) implied volatility surface function (a **smile**)

$$\sigma_{imp}(n, K) = 0.15 + 0.1\left(1 - \frac{K}{90}\right)^2$$

Exercise 11.7. Show that equation (11.31) implies $0 < \pi(n-1, j) < 1$ for $j = 0, 1, 2, ..., n-1$.

Exercise 11.8. Use the forward induction formula for state prices (see Section section 8.7) and the formula (10.1) to show that equation (11.32) holds.

Exercise 11.9. Show that equations (11.32) and (11.33) imply equation (11.34).

Exercise 11.10. Show that equations (11.35) and (11.36) hold.

Exercise 11.11. Show that

$$RS(0,0) = \sum_{j=0}^{n} \lambda(n,j) f(n,j) \tag{11.40}$$

holds for any n.

Exercise 11.12. Show that

$$C(f(n-1,i),n) - P(f(n-1,i),n) = S(0,0) - R^{-n}f(n-1,i) \quad (11.41)$$

holds for any n and i.

Exercise 11.13. Let

$$\Delta^p(n-1,i) = RP(f(n-1,i),n)$$
$$- \sum_{j=0}^{i-1} \lambda(n-1,j)\left[f(n-1,i) - f(n-1,j)\right]$$

and show that $\Delta^p(n-1,i) = \Delta^c(n-1,i)$ for all $i = 0, 1, 2, ..., n-1$.

12
Implied Binomial Trees

The basic references here are Rubinstein [65] and [66]. The latter was reprinted in Jarrow [42]. We shall also provide an extension of the implied binomial tree method by Jackwerth [39]. There are several other extensions that we do not treat, for example, Brown and Toft [10] and Lim and Zhi [46].

The goal here is to produce a binomial (recombining) tree model for a "stock" (or currency, etc.), which will produce prices for some options that agree with market observed prices. Once obtained, the applications are similar to those described in Chapter 11. Once we have completed our discussion we may wish to compare our results with the Derman and Kani approach of the previous chapter.

12.1 The Inputs

The inputs for this construction are the following:

S, the **present value** of the "stock" price.

An **expiry date** T. We will write $T = N \cdot \Delta t$, where N will be specified below.

Interest rates. Rubinstein assumes constant interest rates so that (in our notation) $R(n,j) = R = \exp(r \cdot \Delta t)$ for all (n,j). It will be obvious from the procedure that the method can be generalized to the case $R(n,j) = R(n) = \exp(r(n) \cdot \Delta t)$ for all (n,j) for a given set of values for $r(n)$, $n = 0, 1, 2, ...$

Dividends. Lumpy or continuous dividends can be incorporated, but we shall not discuss dividends in these notes. The case of continuous (proportional), dividends can be more easily incorporated by replacing $R(n,j)$ above with $R(n,j) = R(n) = \exp((r(n) - \delta(n)) \cdot \Delta t)$, for suitable $\delta(n)$. If the "stock" is a currency (exchange rate), then $r(n) = r_d(n)$, the domestic interest rate, and $\delta(n) = r_f(n)$, the foreign interest rate. Even though this case is easy to incorporate, we shall leave details to the reader.

The **possible stock values** at time T will be $S(N,j)$, for $j = 0,1,2,..,N$. There are various way these can be chosen. We shall describe two methods: **Rubinstein's 1994** and **van der Hoek's 1998**.

European call market prices. These all have expiry at time T. Their present values will be denoted by $\tilde{C}_i$ and will correspond to strike prices K_i for $i = 1,2,3,\ldots,m$. These will be known market prices, and we shall construct the binomial tree so that the "theoretical" values from our model (C_i) give these market values: $C_i = \tilde{C}_i$. **Rubinstein** relaxes this to the requirement that $C_i^b \leq C_i \leq C_i^a$, where C^a and C^b are (market) bid and ask prices for calls. He then also relaxes $S(0,0) = S$ to $S^b \leq S(0,0) \leq S^b$. While these features can be included we shall require that $S(0,0) = S$ and $C_i = \tilde{C}_i$.

12.2 Time T Risk-Neutral Probabilities

Once the tree has been constructed we shall show

$$S = S(0,0) = \sum_{j=0}^{N} \lambda(N,j) S(N,j) \tag{12.1}$$

and

$$\tilde{C}_i = C_i = \sum_{j=0}^{N} \lambda(N,j) \left[S(N,j) - K_i \right]^+ \text{ for } i = 1,2,\ldots,m. \tag{12.2}$$

We shall need to compute the values for $\lambda(N,j)$, or equivalently Q_j, where

$$\lambda(N,j) = \frac{Q_j}{R^N}, \quad j = 0,1,\ldots N. \tag{12.3}$$

Here $Q_j > 0$ and $Q_0 + Q_1 + \ldots + Q_N = 1$. The Q_j will be called the **time T risk-neutral probabilities**. We shall give methods for their computation below. For now we assume that we know $\{S(N,j), Q_j\}$ for $j = 0,1,\ldots,N$. In summary we seek $Q_j > 0$, $j = 0,1,\ldots,N$ so that

$$1 = \sum_{j=0}^{N} Q_j \tag{12.4}$$

$$R^N \cdot S = \sum_{j=0}^{N} Q_j S(N,j) \tag{12.5}$$

$$R^N \cdot \tilde{C}_i = \sum_{j=0}^{N} Q_j \left[S(N,j) - K_i\right]^+ \quad \text{for } i = 1, 2, \ldots, m. \tag{12.6}$$

Remark 12.1. An interesting issue is whether there exist choices of $S(N,j)$ and Q_j for $j = 0, 1, 2, \ldots, N$ so that (12.4), (12.5), (12.6) hold. If $N \geq m$ this is possible (as we shall see).

12.3 Constructing the Binomial Tree

We now describe the **As Simple as One, Two, Three** construction. We define $q(n,j)$ as the risk-neutral probability of a path ending at (n,j) at time $t = n$. We have the recurrence

$$q(n+1, j+1) = q(n,j) \cdot \pi(n,j) \tag{12.7}$$

$$q(n+1, j) = q(n,j) \cdot [1 - \pi(n,j)] \tag{12.8}$$

$$q(n+1, 0) = q(n,0) \cdot [1 - \pi(n,0)]. \tag{12.9}$$

We assume that $q(n,j)$ is independent of the path that leads from $(0,0)$ to (n,j). With $C_j^n = \frac{n!}{j!(n-j)!}$, define

$$Q(n,j) = C_j^n \cdot q(n,j). \tag{12.10}$$

This gives the risk-neutral probability of ending in state j at time $t = n$. Recall that in our recombining binomial tree there are C_j^n paths leading $(0,0)$ to (n,j), each with the same risk-neutral probability $q(n,j)$. Then

$$Q(N,j) = Q_j \quad \text{for } j = 0, 1, 2, \ldots, N. \tag{12.11}$$

So

$$q(N,j) = \frac{Q_j}{C_j^N} \quad \text{for } j = 0, 1, 2, \ldots, N. \tag{12.12}$$

Now by (12.7) and (12.8) we backwardize

$$q(n,j) = q(n+1, j+1) + q(n+1, j) \tag{12.13}$$

so we compute

$$\pi(n,j) = \frac{q(n+1, j+1)}{q(n,j)} = \frac{q(n+1, j+1)}{q(n+1, j+1) + q(n+1, j)} \tag{12.14}$$

and

$$S(n,j) = \frac{\pi(n,j)S(n+1, j+1) + (1 - \pi(n,j))S(n+1, j)}{R}. \tag{12.15}$$

Here here we could have used $R(n)$ in place of R.

One, Two, Three refers to equations (12.13), (12.14), (12.15).

Example 12.2. This is part of an example from Rubinstein.
$N = 3$ and $S = 100$. $S(3,0) = 78.27$, $S(3,1) = 92.16$, $S(3,2) = 108.51$, $S(3,3) = 127.76$. $Q_0 = 0.10$, $Q_1 = 0.40$, $Q_2 = 0.30$, $Q_3 = 0.20$,

Note that (12.5) holds.

$$\sum_j Q_j S(3,j) = 102.796 = 100 \cdot R^3$$

implies $R = 1.009234462...$ Actually R will be chosen rather than calculated. Then

$$q(3,0) = \frac{Q_0}{C_0^3} = \frac{Q_0}{1} = 0.1000$$

$$q(3,1) = \frac{Q_1}{C_1^3} = \frac{Q_1}{3} = 0.1333$$

$$q(3,2) = \frac{Q_2}{C_2^3} = \frac{Q_2}{3} = 0.1000$$

$$q(3,3) = \frac{Q_3}{C_3^3} = \frac{Q_3}{1} = 0.2000.$$

We shall perform all calculations to four decimal places. Then

$$q(2,0) = q(3,0) + q(3,1) = 0.2333$$

12.3 Constructing the Binomial Tree

$$q(2,1) = q(3,1) + q(3,2) = 0.2333$$

$$q(2,2) = q(3,2) + q(3,3) = 0.3333$$

and (for interest)

$$Q(2,0) = 1 \cdot q(3,0) = 0.2333$$

$$Q(2,1) = 2 \cdot q(3,1) = 0.4666$$

$$Q(2,2) = 1 \cdot q(3,2) = 0.3333,$$

which sum to 1. Then

$$\pi(2,0) = \frac{q(3,1)}{q(2,0)} = 0.5714$$

$$\pi(2,1) = \frac{q(3,2)}{q(2,1)} = 0.4286$$

$$\pi(2,2) = \frac{q(3,3)}{q(2,2)} = 0.6667.$$

Further

$$S(2,0) = \frac{\pi(2,0)S(3,1) + (1-\pi(2,0))S(3,0)}{R} = 85.4184$$

$$S(2,1) = \frac{\pi(2,1)S(3,2) + (1-\pi(2,1))S(3,1)}{R} = 98.2598$$

$$S(2,2) = \frac{\pi(2,2)S(3,3) + (1-\pi(2,2))S(3,2)}{R} = 120.2330.$$

So that completes steps One, Two, Three for $t = 2$. Now

$$q(1,0) = q(2,0) + q(2,1) = 0.4667$$
$$q(1,1) = q(2,1) + q(2,2) = 0.5333,$$

and for interest,

$$Q(1,0) = 1 \cdot q(1,0) = 0.4667$$
$$Q(1,1) = 1 \cdot q(1,1) = 0.5333,$$

whose sum is 1. Then

$$\pi(1,0) = \frac{q(2,1)}{q(1,0)} = 0.5000$$
$$\pi(1,1) = \frac{q(2,2)}{q(1,1)} = 0.5625$$

and

$$S(1,0) = \frac{\pi(1,0)S(2,1) + (1 - \pi(1,0))S(2,0)}{R} = 90.9987$$

$$S(1,1) = \frac{\pi(1,1)S(2,2) + (1 - \pi(1,1))S(2,1)}{R} = 109.6076$$

So that completes steps One, Two, Three for $t = 1$. Finally,

$$q(0,0) = q(1,1) + q(1,0) = 1.0000$$

$$Q(0,0) = 1 \cdot q(0,0) = 1.0000$$

$$\pi(0,0) = \frac{q(1,1)}{q(0,0)} = 0.5333$$

$$S(0,0) = \frac{\pi(0,0)S(1,1) + (1 - \pi(0,0))S(1,0)}{R} = 100.0000$$

and we are complete. Note that $\pi(n,j)$ is not constant as in the CRR model.

12.4 A Basic Theorem and Applications

Theorem 12.3. *Suppose that*

$$V(n,j) = \frac{\pi(n,j)V(n+1,j+1) + (1 - \pi(n,j))V(n+1,j)}{R} \qquad (12.16)$$

for all (n, j). Then

$$J(n) \equiv \sum_{j=0}^{n} Q(n,j) \frac{V(n,j)}{R^n} \qquad (12.17)$$

is independent of n. Recall that here

$$Q(n,j) = C_j^n \cdot q(n,j). \qquad (12.18)$$

Proof. Write

$$Z(n,j) = \frac{V(n,j)}{R^n}$$

then

$$Z(n,j) = \pi(n,j)Z(n+1,j+1) + (1 - \pi(n,j))Z(n+1,j). \qquad (12.19)$$

Therefore,

$$\begin{aligned}
J(n) &= \sum_{j=0}^{n} Q(n,j)Z(n,j) \\
&= \sum_{j=0}^{n} Q(n,j)\left[\pi(n,j)Z(n+1,j+1) + (1 - \pi(n,j))Z(n+1,j)\right] \\
&= \sum_{j=0}^{n} C_j^n \cdot q(n,j)\left[\pi(n,j)Z(n+1,j+1) + (1 - \pi(n,j))Z(n+1,j)\right] \\
&= \sum_{j=0}^{n} C_j^n \left[q(n+1,j+1)Z(n+1,j+1) + q(n+1,j)Z(n+1,j)\right] \\
&= q(n+1, n+1)Z(n+1, n+1) \\
&\quad + \sum_{j=0}^{n-1} C_j^n q(n+1,j+1)Z(n+1,j+1) \\
&\quad + \sum_{j=1}^{n} C_j^n q(n+1,j)Z(n+1,j) + q(n+1,0)Z(n+1,0) \\
&= q(n+1, n+1)Z(n+1, n+1) \\
&\quad + \sum_{j=1}^{n} \left[C_j^n + C_{j-1}^n\right] q(n+1,j)Z(n+1,j) + q(n+1,0)Z(n+1,0)
\end{aligned}$$

$$= \sum_{j=0}^{n+1} C_j^{n+1} q(n+1,j) Z(n+1,j)$$

$$= \sum_{j=0}^{n+1} Q(n+1,j) Z(n+1,j)$$

$$= J(n+1).$$

Here we used

$$C_j^{n+1} = C_j^n + C_{j-1}^n.$$

Thus

$$J(n) = J(n+1)$$

for any n, and so the theorem has been proved. □

Remark 12.4. Equation (12.19) implies that $\{Z(n,\cdot)\}$ is a martingale process. Such processes have constant means. This is what the theorem states.

Example 12.5. Use $V(n,j) = R^n$. Then (12.16) holds trivially. Thus, $J(N) = J(n)$ for any $n < N$. But $J(N) = Q_0 + Q_1 + \ldots + Q_N = 1$, so $J(n) = 1$ for all $n < N$, that is,

$$\sum_{j=0}^{n} Q(n,j) = 1. \tag{12.20}$$

Example 12.6. Use $V(n,j) = S(n,j)$, then (12.16) holds, so again $J(0) = J(N)$ implies

$$Q(0,0)S(0,0) = \sum_{j=0}^{N} Q(N,j) \frac{S(N,j)}{R^N} = \frac{1}{R^N} \sum_{j=0}^{N} Q_j S(N,j) = S \tag{12.21}$$

by (12.5) and so by Example 12.5, (with $n = 0$), $S(0,0) = S$.

Example 12.7. Use $V(n,j) = C_i(n,j)$, which is the value of the European call option priced by the constructed tree with strike price K_i. As this satisfies (12.16), $J(0) = J(N)$ implies

$$Q(0,0)C_i(0,0) = \sum_{j=0}^{N} Q(N,j) \frac{V(N,j)}{R^N}$$

$$= \sum_{j=0}^{N} Q_j \frac{[S(N,j) - K_i]^+}{R^N} = \tilde{C}_i, \qquad (12.22)$$

so by Example 12.5 (with $n = 0$), $C_i(0,0) = \tilde{C}_i$.

Remark 12.8. What these three examples show is that if we construct the binomial tree as described, then the model will satisfy (12.1) and (12.2). This means also that this tree produces the market values for the various call options.

12.5 Choosing Time T Data

We now address the question of choosing $S(N,j)$ and Q_j so that equations (12.4), (12.5) and (12.6) hold.

Rubinstein's 1994 Method

Construct a CRR tree with $S(0,0) = S$ and $N \geq m$ sufficiently large. Then set

$$S(N,j) = S(0,0)u^j d^{N-j}$$

and let

$$Q'_j = C_j^N \pi^j (1-\pi)^{N-j}$$

where $\pi = (R-d)/(u-d)$. Rubinstein then solves the **quadratic linear programming** problem.

Find $Q_j \geq 0 : j = 0, 1, 2, \ldots, N$ with

$$\sum_{j=0}^{N} (Q_j - Q'_j)^2 = \text{minimum} \qquad (12.23)$$

subject to (12.4), (12.5) and (12.6), which we state again:

$$1 = \sum_{j=0}^{N} Q_j \qquad (12.4)$$

$$\exp(rT) \cdot S = \sum_{j=0}^{N} Q_j S(N, j) \qquad (12.5)$$

$$\exp(rT) \cdot \tilde{C}_i = \sum_{j=0}^{N} Q_j \left[S(N, j) - K_i \right]^{+} \quad \text{for } i = 1, 2, \ldots, m. \qquad (12.6)$$

There are various standard routines for solving such problems.

Remark 12.9. Are there feasible solutions? That is, are there $Q_j \geq 0$ which satisfy (12.4),(12.5),(12.6). The answer is "yes" if we assume that there are no arbitrage prices amongst the market prices, the inputs. This is not a trivial result. The details are provided in Appendix D.

If the problem is feasible, then there will be an optimal solution as the minimization is of a continuous function over a bounded and closed set.

We may not be able to guarantee $Q_j > 0$ for all j for the optimal solution. If it turns out that for the optimal solution (which is unique—hence we can say "the") has $Q_{j1} = 0$, then we can drop $S(N, j1)$ and reduce N to $N-1$ and change Δt to $\frac{N}{N-1} \cdot \Delta t$ so R is replaced by $\exp(r \frac{T}{N-1})$. We may need to repeat this argument a number of times. Clearly not all the $Q_j = 0$, so this culling must eventually stop.

Rubinstein [63] discusses other objective functions in place of the one in (12.23).

van der Hoek's 1998 Method

We select the $S(N, j)$ in such a way that a solution for the Q_j can be written down **explicitly**. For simplicity we shall assume that the strike prices are equally spaced, but this can be relaxed. However, the formulae are then a little more complicated [see Appendix E]. In practice the market prices are usually given with equally spaced strike prices. Let $\Delta = K_i - K_{i-1}$ for each $i = 2, 3, \ldots, m$. Some alternative choices are given later.

Set $N = m - 1$, as we shall see. $\rho = \exp(r \cdot T)$ or $P(0, T)^{-1}$.

Choose Q_j as follows:

$$Q_j = \frac{\rho}{\Delta} \left[\tilde{C}_{j-1} - 2\tilde{C}_j + \tilde{C}_{j+1} \right] \quad \text{for } j = 2, 3, \ldots, m-1 \qquad (12.24)$$

$$Q_j = 0 \quad \text{for } j = 1, m \qquad (12.25)$$

$$Q_{m+1} = \frac{\rho}{\Delta} \cdot \left[\tilde{C}_{m-1} - \tilde{C}_m \right] \qquad (12.26)$$

$$Q_0 = 1 - \sum_{j=1}^{m+1} Q_j. \qquad (12.27)$$

12.5 Choosing Time T Data

Choose S_j as follows:

$$S_0 = \rho \cdot \frac{S - \tilde{C}_1 + K_1 \cdot \frac{\tilde{C}_2 - \tilde{C}_1}{\Delta}}{1 + \rho \cdot \frac{\tilde{C}_2 - \tilde{C}_1}{\Delta}} \quad (12.28)$$

$$S_j = K_j \quad \text{for } j = 1, 2, \ldots, m \quad (12.29)$$

$$S_{m+1} = \rho \cdot \frac{\tilde{C}_m}{Q_{m+1}} + S_m. \quad (12.30)$$

Example 12.10. To see how this works we consider a numerical example with $S = 20.14$ and an expiry date of 30 days. Suppose $\rho = 1.0043$. The call prices are $C(18.00) = 2.27$, $C(19.00) = 1.35$, $C(20.00) = 0.63$, $C(21.00) = 0.22$. Thus $\Delta = 1$.

We then calculate by the above formulae the values of Q_j and S_j as seen in Table 12.1.

Table 12.1. Values of Q_j and S_j.

j	S	Q
0	17.3009	0.0760
1	18.0000	0.0000
2	19.0000	0.2009
3	20.0000	0.3113
4	21.0000	0.0000
5	21.5366	0.4118

We leave it to the reader to verify this is a solution. Of course we have rounded off our answers to four decimal places. In this example $m = 4$, as we had 4 option prices. We note that there are 4 nonzero values of Q_j here. So we choose $N = 3$, then we set $S(3, j)$ and $q(3, j)$ as in Table 12.2.

Table 12.2. Values of $S(3, j)$ and $q(3, j)$.

j	$S(3, j)$	$q(3, j)$
0	17.3009	0.0760
1	19.0000	0.0670
2	20.0000	0.1038
3	21.5366	0.4118

We can then proceed to the **One, Two, Three** algorithm to obtain

Table 12.3. Table of π, q.

j	$\pi(2,j)$	$q(2,j)$
0	0.4682	0.1430
1	0.6078	0.1707
2	0.7987	0.5155

Table 12.4. Table of π, q.

j	$\pi(1,j)$	$q(1,j$
0	0.5442	0.3137
1	0.7512	0.6863

Table 12.5. Table of π, q.

j	$\pi(0,j)$	$q(0,j)$
0	0.6863	1.0000

Now we can build the tree to get

$$S(2,j) = \begin{cases} 18.0706 & \text{for } j=0 \\ 19.5798 & \text{for } j=1 \\ 21.1969 & \text{for } j=2. \end{cases}$$

Then

$$S(1,j) = \begin{cases} 18.8649 & \text{for } j=0 \\ 20.7649 & \text{for } j=1. \end{cases}$$

and

$$S(0,0) = 20.1400.$$

12.6 Some Proofs and Discussion

In general a butterfly spread is obtained by buying calls with strikes X_1 and $X_3 > X_1$, and selling two calls with strike $X_2 = (X_1 + X_3)/2$.

We note that $Q_j > 0$ for $j = 2, 3, \ldots, m-1$ because $\tilde{C}_{j-1} - 2\tilde{C}_j + \tilde{C}_{j+1}$ is then the price of a butterfly spread.

Also $Q_{m+1} > 0$ as $\tilde{C}_m < \tilde{C}_{m-1}$, for we have assumed that $K_1 < K_2 < \ldots < K_m$. Furthermore,

$$\sum_{j=1}^{m+1} Q_j = \frac{\rho}{\Delta} \cdot (\tilde{C}_1 - \tilde{C}_2) > 0$$

and

$$\begin{aligned} Q_0 &= 1 - \frac{\rho}{\Delta} \cdot (\tilde{C}_1 - \tilde{C}_2) \\ &> 1 - \frac{\rho}{\Delta} PV(K_2 - K_1) = 0. \end{aligned}$$

Therefore, $Q_0 > 0$

One can show that if $K_2 > K_1$ then $C(K_1) - C(K_2) < PV(K_2 - K_1)$. In fact, suppose the reverse holds, viz., $C(K_1) - C(K_2) \geq PV(K_2 - K_1)$. Then short the K_1 call, buy the K_2 call and invest $PV(K_2 - K_1)$. At expiry date of the calls, the value of this portfolio is

$$\begin{aligned} V(T) &= -(S(T) - K_1)^+ + (S(T) - K_2)^+ + (K_2 - K_1) \\ &= \begin{cases} K_2 - K_1 > 0 & \text{if } S(T) < K_1 \\ K_2 - S(T) > 0 & \text{if } K_1 \leq S(T) < K_2 \\ 0 & \text{if } S(T) \geq K_2, \end{cases} \end{aligned}$$

which gives a type 2 arbitrage opportunity. This again is where we use the assumption that there are no arbitrage opportunities in our models.

We have thus shown that $Q_j > 0$ for $j = 0, 2, 3, \ldots, m-1, m+1$ and their sum is 1.

We now establish

$$\rho S = \sum_{j=0}^{m+1} Q_j S_j \tag{12.31}$$

and

$$\rho \tilde{C}_i = \sum_{j=0}^{m+1} Q_j \left[S_j - K_i \right]^+. \tag{12.32}$$

In fact

$$\sum_{j=0}^{m+1} Q_j S_j = \frac{\rho}{\Delta} \sum_{j=2}^{m-1} \left[\tilde{C}_{j-1} - 2\tilde{C}_j + \tilde{C}_{j+1}\right] K_j$$

$$+ \rho \tilde{C}_m + \frac{\rho}{\Delta}\left[\tilde{C}_{m-1} - \tilde{C}_m\right] K_m + \rho \left[S - \tilde{C}_1 + K_1 \frac{\tilde{C}_2 - \tilde{C}_1}{\Delta}\right]$$

$$= \frac{\rho}{\Delta} \sum_{j=3}^{m-2} \tilde{C}_j [K_{j+1} - 2K_j + K_{j-1}]$$

$$+ \frac{\rho}{\Delta}[\tilde{C}_1 K_2 + \tilde{C}_2 K_3 + \tilde{C}_m K_{m-1} + \tilde{C}_{m-1} K_{m-2} - 2\tilde{C}_2 K_2$$

$$- 2\tilde{C}_{m-1} K_{m-1}] + \rho \tilde{C}_m$$

$$+ \frac{\rho}{\Delta}\left[\tilde{C}_{m-1} - \tilde{C}_m\right] K_m + \rho \left[S - \tilde{C}_1 + K_1 \frac{\tilde{C}_2 - \tilde{C}_1}{\Delta}\right]$$

$$= \frac{\rho}{\Delta}[C_1 K_2 + \tilde{C}_2 K_3 + \tilde{C}_m K_{m-1} + \tilde{C}_{m-1} K_{m-2} - 2\tilde{C}_2 K_2$$

$$- 2\tilde{C}_{m-1} K_{m-1}] + \rho \tilde{C}_m$$

$$+ \frac{\rho}{\Delta}\left[\tilde{C}_{m-1} - \tilde{C}_m\right] K_m + \rho \left[S - \tilde{C}_1 + K_1 \frac{\tilde{C}_2 - \tilde{C}_1}{\Delta}\right]$$

$$= \rho S + \frac{\rho}{\Delta} \tilde{C}_1[K_2 - \Delta - K_1] + \frac{\rho}{\Delta} \tilde{C}_2[K_3 - 2K_2 + K_1]$$

$$+ \frac{\rho}{\Delta} \tilde{C}_{m-1}[K_{m-2} - 2K_m + K_m] + \frac{\rho}{\Delta} \tilde{C}_m[K_{m-1} - \Delta - K_m]$$

$$= \rho S$$

as required. Also, for $i = 1$,

$$\sum_{j=0}^{m+1} Q_j [S_j - K_1]^+ = \sum_{j=1}^{m+1} Q_j [S_j - K_1]$$

$$= \rho S - Q_0 S_0 - K_1 [1 - Q_0]$$

$$= \rho \tilde{C}_1.$$

We used $S_0 \leq K_1$ (since $C_1 \geq S - \frac{K_1}{\rho}$) and

$$Q_0 S_0 = \rho \left[S - \tilde{C}_1 + K_1 \frac{\tilde{C}_2 - \tilde{C}_1}{\Delta}\right]$$

$$Q_0 = 1 + \rho \frac{\tilde{C}_2 - \tilde{C}_1}{\Delta}.$$

For $i > 1$,

$$\sum_{j=0}^{m+1} Q_j [S_j - K_i]^+ = \sum_{j=i+1}^{m+1} Q_j [S_j - K_i]$$

$$= \sum_{j=i+1}^{m-1} Q_j[K_j - K_i] + Q_{m+1}\left[\rho \frac{\tilde{C}_m}{Q_{m+1}} + K_m - K_i\right]$$

$$= \sum_{j=i+1}^{m-1} Q_j[K_j - K_i] + \rho\tilde{C}_m + \frac{\rho}{\Delta}[\tilde{C}_{m-1} - \tilde{C}_m][K_m - K_i]$$

$$= \frac{\rho}{\Delta}\sum_{j=i}^{m-2} \tilde{C}_j[K_{j+1} - K_i] - 2\frac{\rho}{\Delta}\sum_{j=i+1}^{m-1} \tilde{C}_j[K_j - K_i]$$

$$+ \frac{\rho}{\Delta}\sum_{j=i+2}^{m} \tilde{C}_j[K_{j-1} - K_i] + \rho\tilde{C}_m + \frac{\rho}{\Delta}[\tilde{C}_{m-1} - \tilde{C}_m][K_m - K_i]$$

$$= \frac{\rho}{\Delta}\sum_{j=i+2}^{m-2} \tilde{C}_j[K_{j+1} - 2K_j + K_{j-1}] + \frac{\rho}{\Delta}[\tilde{C}_i\Delta + \tilde{C}_{i+1}2\Delta - 2\tilde{C}_{i+1}\Delta]$$

$$+ \frac{\rho}{\Delta}\left[-2\tilde{C}_{m-1}[K_{m-1} - K_i] + \tilde{C}_{m-1}[K_{m-2} - K_i] + \tilde{C}_m[K_{m-1} - K_i]\right]$$

$$+ \rho\tilde{C}_m + \frac{\rho}{\Delta}[\tilde{C}_{m-1} - \tilde{C}_m][K_m - K_i]$$

$$= \rho\tilde{C}_i + \frac{\rho}{\Delta}\tilde{C}_{m-1}[-2K_{m-1} + K_{m-2} + K_m] + \frac{\rho}{\Delta}\tilde{C}_m[K_{m-1} + \Delta - K_m]$$

$$= \rho\tilde{C}_i,$$

as required.

This shows that **van der Hoek's 1998** recipe works.

Example 12.11. Consider $S = 100$, $\sigma = 0.20$, $T = 1$, $r = 0.05$ and calculate Black and Scholes call prices for $K = 75 + 5m$ for $m = 0, 1, \ldots, 10$. Then $C(75) = 28.9744$, $C(80) = 24.5888$, $C(85) = 20.4693$, $C(90) = 16.6994$, $C(95) = 13.3465$, $C(100) = 10.4506$, $C(105) = 8.0214$, $C(110) = 6.0401$, $C(115) = 4.4666$, $C(120) = 3.2475$, $C(125) = 2.3243$.

In Table 12.6, $Q(10, j)$ and $S(10, j)$ are calculated from **van der Hoek's 1998** algorithm. We then calculate $q(10, j) = \frac{Q(10,j)}{C_j^{10}}$ and we are ready to start the **One, Two, Three** algorithm.

$CRR - S(10, j) = 100u^j d^{10-j}$ where $u = \exp(\sigma\sqrt{\Delta t})$ and $d = 1/u$. We note that the CRR prices are more widespread ([53, 188]) than from **van der Hoek's 1998** algorithm ([70, 137]).

Table 12.6. One, Two, Three algorithm.

j	$Q(10,j)$	$S(10,j)$	$q(10,j)$	$CRR - S(10,j)$	BS–Prob
0	0.0779	70.7289	0.0779	53.1286	0.0795
1	0.0559	80.0000	0.0056	60.2924	0.0559
2	0.0735	85.0000	0.0016	68.4222	0.0732
3	0.0876	90.0000	0.0007	77.6482	0.0872
4	0.0961	95.0000	0.0005	88.1182	0.0956
5	0.0981	100.0000	0.0004	100.0000	0.0976
6	0.0942	105.0000	0.0004	113.4839	0.0938
7	0.0857	110.0000	0.0007	128.7860	0.0855
8	0.0745	115.0000	0.0017	146.1515	0.0744
9	0.0622	120.0000	0.0062	165.8584	0.0622
10	0.1941	137.5879	0.1942	188.2227	0.1951

The column BS–Prob lists the Black and Scholes risk-neutral probabilities that $S(T) \in [K - 0.5\Delta, K + 0.5\Delta]$. We use the fact

$$P_{rn}(S(T) > K) = \mathcal{N}(d_2(K)), \tag{12.33}$$

from which we get the approximations

$$P_{rn}(S(T) \in [K - 0.5\Delta, K + 0.5\Delta]) = \mathcal{N}(d_2(K - 0.5\Delta)) - \mathcal{N}(d_2(K + 0.5\Delta))$$
$$\approx \frac{1}{2}[\mathcal{N}(d_2(K - \Delta)) - \mathcal{N}(d_2(K + \Delta))]$$

for $K = K_j$ with $j = 1, 2, \ldots, 11$. Further,

$$P_{rn}(S(T) \in [K_1 - 0.5\Delta, K_1 + 0.5\Delta]) \approx 1 - \frac{1}{2}[\mathcal{N}(d_2(K)) + \mathcal{N}(d_2(K + \Delta))]$$
$$P_{rn}(S(T) \in [K_{12} - 0.5\Delta, K_1 + 0.5\Delta]) \approx \frac{1}{2}[\mathcal{N}(d_2(K)) + \mathcal{N}(d_2(K - \Delta))]$$

We note that $BS - Prob(10,j) \approx Q(10,j)$ for each j. The resulting tree in this example can be used to price American and exotic derivatives under the Black and Scholes model for the evolution of "stock" prices.

12.7 Jackwerth's Extension

The procedure that we have described constructs a binomial tree that prices a number of European call options with the same expiry consistent with market prices. It is possible to modify this construction to consistently make the

12.7 Jackwerth's Extension

tree price some other options (American style options, options with different maturities, path dependent options, etc). Jackwerth [39] showed how to do this with his generalized binomial trees.

Let us first review some earlier calculations. By (12.10) and (12.13)

$$\begin{aligned}
Q(n,j) &= C_j^n q(n,j) \\
&= C_j^n \left[q(n+1,j) + q(n+1,j+1) \right] \\
&= C_j^n \left[\frac{Q(n+1,j)}{C_j^{n+1}} + \frac{Q(n+1,j+1)}{C_{j+1}^{n+1}} \right] \\
&= \omega(n+1,j+1) Q(n+1,j+1) + (1 - \omega(n+1,j)) Q(n+1,j)
\end{aligned}$$
(12.34)

where $\omega(n,j) \equiv \frac{j}{n}$, as can be easily checked.

This suggests a way for generalization

$$Q(n,j) = W\left(\frac{j+1}{n+1}\right) Q(n+1, j+1) + \left(1 - W\left(\frac{j}{n+1}\right)\right) Q(n+1, j) \quad (12.35)$$

where W is any function that satisfies

$$W(0) = 0 \quad (12.36)$$
$$W(1) = 1 \quad (12.37)$$
$$0 < W(\theta) < 1, \quad \text{for } 0 < \theta < 1. \quad (12.38)$$

In the earlier sections we used $W(\theta) \equiv \theta$, but we could choose W to be piecewise linear—for example:

$$W(\theta) = \begin{cases} c\theta & \text{if } 0 \le \theta \le \frac{1}{2} \\ (2-c)\theta + (c-1) & \text{if } \frac{1}{2} \le \theta \le 1, \end{cases} \quad (12.39)$$

with $0 < c < 2$, which is to be chosen (convex for $0 < c < 1$ and concave for $1 < c < 2$). Other forms can be chosen: piecewise linear functions with several kinks, quadratic functions, S-shaped functions, an so on (see [46] for example). With a choice of **weight function** W, equation (12.35) defines the backward recurrence for $Q(n,j)$, where $Q(N,j)$ are still defined as before. We then define

$$\pi(n,j) \equiv W\left(\frac{j+1}{n+1}\right) \frac{Q(n+1, j+1)}{Q(n,j)}. \quad (12.40)$$

170   12 Implied Binomial Trees

This generalizes the case where $W(\theta) \equiv \theta$ and $Q(n,j) = C_j^n q(n,j)$ above. Given $S(N,j)$, this allows us to determine $S(n,j)$ for each $0 \le n < N$ and $j = 0, 1, 2, ..., n-1$.

In fact Theorem 12.3 will still hold using (12.35) and (12.40) as defining $Q(n,j)$ and $\pi(n,j)$.

The tree constructed by this modification is still consistent with (12.4)–(12.6). The free parameter c in this example can be chosen so that the tree correctly prices a further derivative to market values. Whether such a value of $0 < c < 2$ can always be found for the piecewise linear function in (12.39) is a question that is left to the reader. If not, other choices for W could be considered.

The only restrictions on W are those in (12.36)–(12.38), and in particular (12.38) ensures that $Q(n,j) > 0$ for all (n,j).

If we need to build a tree to correctly price an additional $k \ge 1$ derivatives consistently to market values, we could start with choosing a piecewise linear W with k kinks.

12.8 Exercises

Exercise 12.12. Implement Example 12.10 on a spreadsheet.

Exercise 12.13. Implement Example 12.2.

Exercise 12.14. Construct a four-step implied binomial tree ($n = 0, 1, 2, 3, 4$) for the inputs $S(0,0) = 95$, $r = 5\%$, $\Delta t = 0.25$. We are given the following five European call prices for options expiring at $T = 1$: $C(90) = \$18.14$, $C(95) = \$15.04$, $C(100) = \$12.33$, $C(105) = \$10.00$, $C(110) = \$8.02$. [Here, $C(K)$ stands for the European call price expiring at $T = 1$ with strike price K.]

Exercise 12.15. Show that Theorem 12.3 still holds when $\pi(n,j)$ is defined via (12.40) for any choice of W that satisfies (12.36)–(12.38).

13

Interest Rate Models

We continue from the introduction made in Section 3.4 and construct multi-period models based on available market data. The resulting model can be used for interest rate derivative valuation and extended into other binomial models where we wish to perform pricing under stochastic interest rates.

The difficulty with interest rate modelling is the fact that there are many different interest rates. In Section 3.4 we introduced $P(t,T)$, the value at time t of 1 CAD (or one unit of domestic currency, say) at time T. Let us say this in another way: It is the value at time t of a (fixed income) security that pays 1 CAD at time T (under all conditions, or states of the world). The value of the fixed income security is known at expiry (T), but its values are uncertain at times between the present (t, say) and expiry. If one owns such a security, we rely on the counterparty to pay us 1 CAD at time T without fail. Here is the problem. Not all counterparties are equally reliable, (or what is called creditworthy). For a counterparty with high credit rating, as determined by a credit rating agency like Standard and Poors, we would expect a larger value for each $P(t,T)$. There is now a large and growing literature on **credit risk and credit derivatives**.

Interest rate modelling can be specified by modelling various variables.

1. $P(t,T)$.
2. $y(t,T)$ the yield to maturity defined through

$$P(t,T) = \exp\left(-y(t,T) \cdot (T-t)\right)$$
$$y(t,T) = -\frac{1}{T-t} \ln P(t,T).$$

3. $f(t,T)$ the forward rate defined in continuous time through

$$P(t,T) = \exp\left(-\int_t^T f(t,u)du\right)$$

$$f(t,T) = -\frac{\partial}{\partial T} \ln P(t,T).$$

4. $r(t) \equiv f(t,t)$ the spot rate (instantaneous or overnight rate) and

$$P(t,T) = \mathbf{E}\left[\exp\left(-\int_t^T r(u)du\right) \Big| \mathcal{F}_t\right].$$

where $\mathcal{F}_t \equiv \sigma\{r(u) \mid 0 \le u \le t\}$ represents the past history on the spot rates up to the time t (see [29] for more details of this concept).

We can, in principle, start with any of these notions and derive the others. Observed market data are obviously most closely connected with the quantity $P(t,T)$. The modern **Heath-Jarrow-Morton** framework (or HJM) begins with a model for $f(t,T)$. The Ho and Lee model is probably the simplest model of this kind and we shall describe it below.

We shall ultimately be interested in determining values for $R(n,j) = 1 + r(n,j)$ and $\pi(n,j)$ for various nodes in our tree models, for then we can price contingent claims. As a by-product we can then compute values of $P(t,T)$ for various values of (t,T).

One of the first inputs that we need from the market are the values of $P(0,T)$, ideally for all T. There are various approaches depending on the data used.

13.1 $P(0,T)$ from Treasury Data

Consider the first example.

Example 13.1. ([11]) We have three coupon bonds with expiry $T = 1, 2, 3$ with face value $100 paying coupons once a year (starting in one year's time). Their coupon rates and prices are 6% and $99.07, 7% and $100, 8% and $102.62, respectively. We can then determine $P(0,1)$, $P(0,2)$ and $P(0,3)$ by solving

$$\begin{aligned} 99.07 &= 106 P(0,1) \\ 100 &= 7 P(0,1) + 107 P(0,2) \\ 102.62 &= 8 P(0,1) + 8 P(0,2) + 108 P(0,3) \end{aligned} \tag{13.1}$$

yielding $P(0,1) = 0.9346$, $P(0,2) = 0.8735$ and $P(0,3) = 0.8162$.

Unfortunately, life is not so simple.

Treasury notes/bills have maturities $T < 1$. In the USA they have face value $100 and are classified as **discount securities**. If we have a 90-day bill with 8% discount rate then its price would be

$$98 = 100\left[1 - \frac{1}{4} \times 0.08\right]$$

so we could deduce that $P(0, 0.25) = 0.98$. See Hull [ibid., section 4.4] for further discussion.

There are coupon bonds with coupons paid (usually) each 6 months and often at some standard proportion of the face value (like 6% or 8%). Instead of quoting prices of such bonds, their yields to maturity are often quoted, from which their price can be calculated. Vice versa, the yield to maturity can be uniquely determined from a bond's price. We give these details later. Let us assume that a bond that expires at T_n has coupon payments of C at $T_1, T_2, \ldots, T_n$. It has present value

$$P = \sum_{i=1}^{n} C\, P(0, T_i) + F\, P(0, T_n) \tag{13.2}$$

where F is the face value of the bond, and C is a given percentage (as specified in Federal Reserve documentation) of F.

Let us return to Example 13.1. Suppose that instead of the data about the 3-year bond we have a 4-year bond with value \$105 and coupon rate 9%. Then we need to solve:

$$\begin{aligned} 99.07 &= 106P(0,1) \\ 100 &= 7P(0,1) + 107P(0,2) \\ 105 &= 9P(0,1) + 9P(0,2) + 9P(0,3) + 109P(0,4), \end{aligned} \tag{13.3}$$

but now we can no longer determine (uniquely) the value of $P(0, 3)$ and $P(0, 4)$. We can proceed in various ways.

We could assume an analytic form for $P(0, t)$ like

$$P(0, t) = m(t) = a_0 + b_0 t + c_0 t^2 + d_0 t^3 \tag{13.4}$$

with $a_0 = 1$ as $P(0, 0) = 1$. The equations (13.3) can then be used to compute a_0, b_0, c_0, d_0, and then we obtain other values of $P(0, t)$. Let us note a couple of things. We cannot calculate values of $P(0, t)$ for t greater than the maturity of the bond with longest duration. We know that $P(0, t)$ should be a strictly decreasing function of t; it could happen that solving for the unknown constants in (13.4) leads to a contradiction of this fact.

We could decide to be parsimonious and select a simpler form for $P(0, t)$, like

$$P(0,t) = m(t) = 1 + b_0 t + c_0 t^2 \qquad (13.5)$$

but then (13.3) is overdetermined for (13.5). We could instead seek b_0 and c_0 to minimize

$$J = \sum_{i=1}^{3} w_i \epsilon_i^2 \qquad (13.6)$$

where w_1, w_2, w_3 are positive weights summing to 1, and

$$\epsilon_1 = 99.07 - 106 m(1)$$
$$\epsilon_2 = 100 - 7 m(1) + 107 m(2)$$
$$\epsilon_3 = 105 - 9 m(1) + 9 m(2) + 9 m(3) + 109 m(4).$$

We might even even impose an additional constraint on the parameters so that $m'(t) \leq 0$ for $0 \leq t \leq 4$ ($b_0 \leq 0$ and $b_0 + 8c_0 \leq 0$). These type of curve-fitting issues also arise in various applications in statistics.

If we have bonds out to a long time horizon, like 10 or more years, we could also consider fitting cubic splines in much the same way as we have just described. This means taking

$$m(t) = a_0 + b_0 t + c_0 t^2 + d_0 t^3 + \sum_{j=1}^{k} F_j (t - t_j)_+^3 \qquad (13.7)$$

for "knots" $t_1, t_2, \ldots, t_k$. We now need to find $(k+3)$ unknowns (recall $a_0 = 1$) to fit the bond data exactly or with a weighted least squares method. Usually only a few knots are used, placing most in the near time zone [0,2], where most of the "action" is.

Some references to these methods include McCulloch [52] and Litzenberger and Rolfo [47]. See also Appendix F. It seems that many practitioners calculate the treasury's discount curve as a spread to the discount curve from BBSW data, which we describe in the next section.

13.2 $P(0,T)$ from Bank Data

We are now talking about the AA market. This is what we shall call the market that quotes the bank bill swap rates, which are only approximately AA in credit rating. We no longer have bonds, but the banks use what are called **bank bill swap (BBSW) rates**. We now present some of these details.

Swap Rates

The variable rate that applies between times t and $t + \delta$ will be r_t (or more precisely, $r_{t,t+\delta}$). This rate will be known with certainty at time t. Investing \$1 at time t will yield $1 + r_t \delta$ at time $t + \delta$. Now as

$$(1 + r_t \delta) P(t, t + \delta) = 1 \qquad (13.8)$$

we have

$$r_t \equiv r_{t,t+\delta} = \frac{1}{\delta} \left[\frac{1}{P(t, t+\delta)} - 1 \right]. \qquad (13.9)$$

Consider now a swap performed at time $t + \delta$. A **payer-swap** is one in which a fixed rate is paid and a floating/variable rate is received. A **receiver-swap** has the reverse cash flow. If the face value (or principle) is F and the fixed rate is κ, then the net flow is

$$F \cdot r_t \cdot \delta - F \cdot \kappa \cdot \delta. \qquad (13.10)$$

This can also be written as

$$F \cdot (r_t - \kappa) \cdot \delta = F \cdot (1 + r_t \delta) - F \cdot (1 + \kappa \delta)$$
$$= F \cdot \frac{1}{P(t, t+\delta)} - F \cdot (1 + \kappa \delta).$$

This has present value (at time 0) given by

$$PV = F \cdot P(0, t) - F \cdot (1 + \kappa \delta) \cdot P(0, t + \delta)$$
$$= F \cdot [P(0, t) - P(0, t + \delta)] - F \cdot \kappa \cdot \delta \cdot P(0, t + \delta). \qquad (13.11)$$

Now consider a (payer-) swap where interest is exchanged at times $\delta, 2\delta \ldots, n\delta$. The present value of all these cash flows together is (using (13.11) repeatedly):

$$PV = F \cdot [P(0, 0) - P(0, \delta)] - F \cdot \kappa \cdot \delta \cdot P(0, \delta)$$
$$+ F \cdot [P(0, \delta) - P(0, 2\delta)] - F \cdot \kappa \cdot \delta \cdot P(0, 2\delta)$$
$$+ F \cdot [P(0, 2\delta) - P(0, 3\delta)] - F \cdot \kappa \cdot \delta \cdot P(0, 3\delta)$$
$$+ \cdots$$
$$+ F \cdot [P(0, (n-1)\delta) - P(0, n\delta)] - F \cdot \kappa \cdot \delta \cdot P(0, n\delta)$$

$$= F \cdot [1 - P(0, n\delta)] - F \cdot \kappa \cdot \delta \cdot [P(0, \delta) + P(0, 2\delta) + \ldots + P(0, n\delta)]. \tag{13.12}$$

The choice of κ, the swap rate, is made so that the expression in (13.12) is 0. Thus, we obtain the basic formula for κ:

$$\kappa = \frac{1}{\delta} \left[\frac{1 - P(0, n\delta)}{P(0, \delta) + P(0, 2\delta) + \ldots + P(0, n\delta)} \right]. \tag{13.13}$$

For semiannual swaps, $\delta = 0.5$, and for quarterly swaps, $\delta = 0.25$. We shall write $\kappa(nQ)$ for the n-year quarterly swap rate, and $\kappa(nS)$ for the n-year semiannual swap rate.

Computing P(0,t) from Market Data

The AA market provides BBSW (bank bill swap rates) for 30, 60, 90, 120, 150 and 180 days. Then $P(0,t)$ for $t = \frac{1}{12} = \frac{30}{360}, \frac{2}{12}, \ldots, \frac{6}{12}$ are given by the simple discounting formula. For example,

$$P(0, 0.25) = P(0, \frac{3}{12}) = \frac{1}{1 + \frac{90}{360} \cdot i_{90}} \tag{13.14}$$

where i_{90} is the 90-day BBSW interest rate. Data and calculations are provided for an example with rates (quoted as percents) as of closing for a date in 2002: 4.870, 4.910, 4.925, 4.910, 4.940, 4.950, and the overnight rate was 4.750.

The AA market provides 1Q, 2Q, 3Q (Q for quarterly) swap rates and 4S, 5S, 7S, 10S, 15S and 20S (S for semiannual) swap rates. Examples of such data provided on the same date were 5.000, 5.125, 5.415 and 5.580, 5.680, 5.855, 6.010, 6.110, 6.160. Such data are provided by Reuters (for example).

Practitioners like to use semiannual swap rates and they often use the approximation

$$\left[1 + \frac{\kappa(nQ)}{4}\right]^4 \approx \left[1 + \frac{\kappa(nS)}{2}\right]^2$$

or

$$\kappa(nS) = 2\left[\left[1 + \frac{\kappa(nQ)}{4}\right]^2 - 1\right] \tag{13.15}$$

when quarterly swap data are given $(1Q, 2Q, 3Q)$. The output from (13.15) is illustrated in Table 13.1.

Whether (13.15) is a good approximation can only be tested in any particular interest rate model. It seems to work well in the Hull-White one-factor interest

13.2 $P(0,T)$ from Bank Data

Table 13.1. Discounting functions.

Data 8/12/2002				$P(0,t)$	z	a	b	c
	overnight	4.750		1				
BBSW	30	4.870		0.9960				
	60	4.910		0.9919				
	90	4.925		0.9878				
	120	4.910		0.9839				
	150	4.940		0.9798				
	180	4.950		0.9758				
				(approx)				
SWAPS	1Q	5.000	1S	5.031	0.9515	0.97423	0.97605	0.0245 -0.95029
				0.9270				
	2Q	5.125	2S	5.158	0.9031	0.97045	0.92772	0.0246 -0.89758
				0.8764				
	3Q	5.415	3S	5.452	0.8764	0.95575	0.90086	0.0245 -0.84626
				0.8376				
	4S	5.580	4S	5.580	0.8006	0.97018	0.82329	0.0227 -0.79698
				0.7767				
	5S	5.680	5S	5.680	0.7535			
	7S	5.855	7S	5.855				
	10S	6.010	10S	6.010				
	15S	6.110	15S	6.110				
	20S	6.160	20S	6.160				

rate model—this is an example of an interest rate model with explicit formulas for zero-coupon bond prices.

With this approximation the approximate values for $\kappa(1S)$, $\kappa(2S)$ and $\kappa(3S)$ are 5.041, 5.272 and 5.487 respectively.

Given $\kappa(1S)$ (or rather its approximation) and $P(0,0.5)$ we can compute $P(0,1)$ from

$$\kappa(1S) = 2\left[\frac{1-P(0,1)}{P(0,0.5)+P(0,1)}\right].$$

This gives

$$P(0,1) = \frac{1 - \frac{\kappa(1S)}{2} \cdot P(0,0.5)}{1 + \frac{\kappa(1S)}{2}}. \tag{13.16}$$

We now compute $P(0,1.5)$ and $P(0,2)$. Recall

$$\kappa(2S) = 2\left[\frac{1-P(0,2)}{P(0,0.5)+P(0,1)+P(0,1.5)+P(0,2)}\right],$$

which is one equation for two unknowns $P(0,1.5)$ and $P(0,2)$. The way that practitioners deal with this is to assume that for some (forward) rate y we have

$$P(0,1.5) = P(0,1) \cdot \left[\frac{1}{1+\frac{y}{2}}\right]$$

$$P(0,2) = P(0,1) \cdot \left[\frac{1}{1+\frac{y}{2}}\right]^2.$$

Setting

$$z = \frac{1}{1+\frac{y}{2}}$$

we have a quadratic equation for z:

$$az^2 + bz + c = 0. \tag{13.17}$$

Here

$$a = P(0,1) \cdot \left[1 + \frac{\kappa(2S)}{2}\right]$$

$$b = P(0,1) \cdot \left[\frac{\kappa(2S)}{2}\right]$$

$$c = \frac{\kappa(2S)}{2} \cdot [P(0,0.5) + P(0,1)] - 1$$

We then set

$$P(0,1.5) = z \cdot P(0,1)$$
$$P(0,2) = z^2 \cdot P(0,1),$$

where

$$z = \frac{-b + \sqrt{b^2 - 4 \cdot a \cdot c}}{2 \cdot a}.$$

This procedure can be generalized when values of $\kappa(nS)$ are given. In fact, if

$$P(0,0.5), P(0,1), P(0,1.5), \ldots, P(0,n-1)$$

are known, then we can compute (estimate) $P(0, n-\frac{1}{2})$ and $P(0,n)$ as follows. Regard y as a forward rate applying over $[n-1, n]$ and suppose z is defined as above. Using the formula for $\kappa(nS)$ we have

$$az^2 + bz + c = 0, \tag{13.18}$$

where

$$a = P(0, n-1) \cdot \left[1 + \frac{\kappa(nS)}{2}\right]$$

$$b = P(0, n-1) \cdot \left[\frac{\kappa(nS)}{2}\right]$$

$$c = \frac{\kappa(nS)}{2} \cdot [P(0, 0.5) + P(0, 1) + \ldots + P(0, n-1)] - 1$$

$$= \frac{\kappa(nS)}{\kappa((n-1)S)} [1 - P(0, n-1)] - 1.$$

Then set

$$P(0, n - \tfrac{1}{2}) = z \cdot P(0, n-1)$$

$$P(0, n) = z^2 \cdot P(0, n-1)$$

where as before

$$z = \frac{-b + \sqrt{b^2 - 4 \cdot a \cdot c}}{2 \cdot a}.$$

We can now give values for $P(0, t)$ for t = 1, 1.5, 2, 2.5, 3, 3.5, 4, 4.5, 5. These are 0.9514, 0.9428, 0.9259, 0.9175, 0.9010, 0.8923, 0.8750, 0.8665 and 0.8497. It is a good exercise to check these computations.

These calculations give values for a period of 5 years. As we do not have values for $\kappa(6S)$, $\kappa(8S)$, $\kappa(9S)$, $\kappa(11S)$ and so on, we need to use further approximations to compute further values of $P(0, t)$ as above. We shall not go into the details of this here. We could generalize the procedure above, but we do not get a quadratic equation for z in that case. However, a simple approach could be obtained by assuming that the longer term rates are "flat" and we obtain the "missing" swap rates by interpolation. Thus, we could approximate

$$\kappa(6S) = \frac{1}{2} \cdot \kappa(5S) + \frac{1}{2} \cdot \kappa(7S)$$

$$\kappa(8S) = \frac{2}{3} \cdot \kappa(7S) + \frac{1}{3} \cdot \kappa(10S) \qquad (13.19)$$

and so forth, and then use the "quadratic" algorithm as above.

Once we have the values of $P(0, t)$ for t a multiple of a half-year, we obtain other values by interpolation.

Computing P(0, T) from Market Data—A Second Version

This is another version used by practitioners. Some treasurers use a simpler approach than the "quadratic method" just described. The procedure is the same as just described for computing $\kappa(nS)$ for integer values of n. The new feature of this method is to set

$$\kappa((n+0.5)S) = 0.5(\kappa(nS) + \kappa((n+1)S)). \tag{13.20}$$

Once these values are known, we can calculate $P(0, n+0.5)$ in terms of $P(0, n)$ whenever n is a multiple of 0.5. We use

$$\kappa((n+0.5)S) = 2\left[\frac{1 - P(0, n+0.5)}{\frac{2(1 - P(0,n))}{\kappa(nS)} + P(0, n+0.5)}\right]. \tag{13.21}$$

This can be arranged to give

$$P(0, n+0.5) = \left[1 + \frac{\kappa((n+0.5)S)}{2}\right]^{-1}\left[1 - \frac{\kappa((n+0.5)S)}{\kappa(nS)}(1 - P(0,n))\right] \tag{13.22}$$

which is simple recursive relation.

Remark 13.2. As a by-product, note that the following must hold:

$$\kappa((n+0.5)S) < \frac{\kappa(nS)}{1 - P(0,n)}. \tag{13.23}$$

Interpolation of P(0, T)

Suppose that we know $P(0, T_1)$ and $P(0, T_2)$, and we wish to estimate a value for $P(0, T)$, where $T_1 < T < T_2$. The following estimate is frequently used:

$$P(0, T) = P(0, T_1)^{1-\alpha} P(0, T_2)^\alpha \tag{13.24}$$

where

$$\alpha = \frac{T - T_1}{T_2 - T_1}. \tag{13.25}$$

This is motivated, rather than proved, as follows. Write

$$P(0, T_2) = P(0, T_1)e^{-\bar{r}(T_2 - T_1)}. \tag{13.26}$$

13.2 $P(0,T)$ from Bank Data

This defines a "forward" rate $\bar{r}$ for the period $[T_1, T_2]$. We then make the approximation

$$P(0,T) = P(0,T_1)e^{-\bar{r}(T-T_1)} \tag{13.27}$$

and so

$$P(0,T) = P(0,T_1)\left[\frac{P(0,T_2)}{P(0,T_1)}\right]^{\frac{T-T_1}{T_2-T_1}}$$

$$= P(0,T_1)^{1-\frac{T-T_1}{T_2-T_1}} P(0,T_2)^{\frac{T-T_1}{T_2-T_1}},$$

as claimed. These interpolated values of $P(0,t)$ can be used to discount certain cash flows that fall on dates which are not integer multiples of 0.5 years ahead.

Other Discount Curves

We can now compute $\tilde{P}(0,t)$ for less creditworthy entities. We then have $\tilde{P}(0,t) < P(0,t)$ for each t, but of course $\tilde{P}(0,0) = P(0,0) = 1$. If we are considering the AA curve, then what we will say still applies, but we will see that $\tilde{P}(0,t) > P(0,t)$ if $P(0,t)$ stands for AA values. In fact we will consider this last situation using hypothetical government (or semigovernment) bond data. We first use market data to compute the discount functions from the AA (say) data as explained above.

Since the credit rating of the government is greater than AA we will expect $\tilde{P}(0,t) > P(0,t)$, as we have already noted. Suppose we have data for two government bonds. One expires on 15 August 2005 and pays a coupon semi-annually at a rate of $c_1 = 6.5\%$ per annum. This means that if the face value of this bond were \$100 you would receive a coupon payment of $(100 \times 0.065)/2 = \$3.75$ every six months. Rather than quoting the value of this bond, it is usual (as we mentioned earlier) to quote the yield (to maturity), which we will call $y_1\%$. The value of this bond (per face value 1) is given by (see [71])

$$MS1 = \frac{c_1}{2}\left[\frac{1}{(1+\frac{y_1}{2})} + \frac{1}{(1+\frac{y_1}{2})^2} + \ldots + \frac{1}{(1+\frac{y_1}{2})^{m_1}}\right] + \frac{1}{(1+\frac{y_1}{2})^{m_1}} \tag{13.28}$$

where $c_1 = 0.065$, $m_1 = 6$ is the number of coupons paid, $y_1 = 0.0537$. The answer is $MS1 = \$1.0309$. It is well known that the expression for MS1 can be simplified (using properties of geometric series) to

$$MS1 = \frac{c_1}{y_1}\left[1 - \left(1+\frac{y_1}{2}\right)^{-m_1}\right] + \left(1+\frac{y_1}{2}\right)^{-m_1}. \tag{13.29}$$

Suppose a second government bond matures on 15 October 2007. The coupon rate $c_2 = 0.075$ or 7.5% again with semi-annual payments. The first coupon after 15 August 2002 will be on 15 October 2002 and every six months after that until expiry. Let τ be the time to the first coupon payment after 15 August 2002. Thus $\tau = 1/6$. In this case $y_2 = 0.05615$ and $m_2 = 11$, and after some computations,

$$MS2 = \frac{c_2}{y_2}\left(1+\frac{y_2}{2}\right)^{1-2\tau}\left[1-\left(1+\frac{y_2}{2}\right)^{-m_2}\right] + \left(1+\frac{y_2}{2}\right)^{-m_2+1-2\tau}. \quad (13.30)$$

One could interpret this formula and discuss how it could be obtained from an expression like (13.29). Note that in the case $\tau = 0.5$, then (13.29) and (13.30) agree. In any case the given data gives $MS2 = \$1.1084$. We shall now assume that

$$\tilde{P}(0,t) = \begin{cases} P(0,t) - x_1 t & \text{if } 0 \leq t \leq t_1 \\ P(0,t) - x_1 t_1 - x_2(t-t_2) & \text{if } t_1 \leq t \leq t_2. \end{cases} \quad (13.31)$$

where $t_1 = 0.5m_1$, the time to expiry of the first bond, and $t_2 = \tau + 0.5(m_2-1)$ the time to expiry of the second bond. We now compute x_1 and x_2 so that

$$MS1 = \frac{c_1}{2}\left[\tilde{P}(0,0.5) + \tilde{P}(0,1) + \cdots \right.$$
$$\left. + \tilde{P}(0,0.5m_1)\right] + \tilde{P}(0,0.5m_1) \quad (13.32)$$

$$MS2 = \frac{c_2}{2}\left[\tilde{P}(0,\tau) + \tilde{P}(0,\tau+0.5) + \cdots \right.$$
$$\left. + \tilde{P}(0,\tau+0.5(m_2-1))\right] + \tilde{P}(0,\tau+0.5(m_2-1)).(13.33)$$

The answers are $x_1 = -0.00095$ and $x_2 = 0.00011$. This leads us to $\tilde{P}(0.t) > P(0,t)$ as we expected.

Once we have then computed the values of $\tilde{P}(0,t)$ we can calculate the adjusted swap rates:

$$\tilde{\kappa}((n+0.5)S) = 2\left[\frac{1-\tilde{P}(0.n+0.5)}{\frac{2(1-\tilde{P}(0.n))}{\tilde{\kappa}(nS)} + \tilde{P}(0,n+0.5)}\right] \quad (13.34)$$

for $n = 0, 0.5, 1, 1.5, \ldots$ using $\tilde{P}(0,0) = 1$. This recurrence is again easy to implement, and the spread $\kappa(nS) - \tilde{\kappa}(nS)$ can be calculated. In this example the spread ranged between 7.5 and 11 basis points (1 basis point is 0.01%).

This describes an actual method used by practitioners [1]. It is tractable, that is, easy to implement.

As we said, one can always study specific interest rate models and decide whether these approximations are in fact accurate.

Nelson-Siegel Type Approach

This approach was introduced by Nelson and Siegel [58]. See also de la Grandville [23].

To explain, we introduce forward rates $f(0, u)$ by

$$P(0,T) = \exp\left(-\int_0^T f(0,u)\,du\right). \tag{13.35}$$

Nelson and Siegel assume that

$$f(0,u) = a + be^{-\alpha u} + cue^{-\alpha u}$$

for some choices of (a, b, c, α). In that case

$$P(0,T) = \exp\left(-A - B \cdot T - (A + C \cdot T)e^{-\alpha T}\right) \tag{13.36}$$

where

$$A = \frac{b}{\alpha} + \frac{c}{\alpha^2} \qquad B = a \qquad C = \frac{c}{\alpha}.$$

We could also set $f(0,0) = a+b$ to be equal to overnight rate (at call). In that case there are only three parameters b, c, α. Write $\theta = (b, c, \alpha)$. Also $\theta_1 \equiv b$, etc. The equation for $\kappa(nS)$ can be written

$$\frac{\kappa(nS)}{2}[P(0,0.5) + \ldots + P(0,n)] - 1 + P(0,n) = 0,$$

and if we substitute (13.35) we get an equation of the form

$$\Phi(\theta) = 0.$$

In fact, we would have a number of such equations

---

[1] The authors would like to acknowledge Bob Arnold of SAFA (South Australian Government Financing Authority) for providing the ideas used here to compute $P(0,t)$ from BBSW data.

$$\Phi_j(\theta) = 0, \quad j = 1, 2, \ldots, m \tag{13.37}$$

and we may choose θ to minimize a weighted square sum

$$J(\theta) \equiv \sum_{j=1}^{m} w_j \Phi_j(\theta)^2$$

where all the w_j are positive and sum to 1. After all, it may not be possible to find a solution for all the equations in (13.37). We could then select θ so that

$$\frac{\partial J(\theta)}{\partial \theta_i} = 0, \quad i = 1, 2, 3,$$

three equations in three unknowns. This is not trivial to solve and requires some numerical analysis and a computer algorithm.

Other families similar to the Nelson-Siegel family could be used. In fact there is much interest in the so-called **consistent families** (consistent with particular interest rate models). See Bjork and Christensen [6]. The following family is consistent with the so-called **Hull-White** interest rate model:

$$f(0, u) = ae^{-\alpha u} + be^{-2\alpha u}$$

for which

$$P(0, T) = \exp\left[\frac{1}{\alpha}\left(ae^{-\alpha T} + \frac{b}{2}e^{-2\alpha T} - \left(a + \frac{b}{2}\right)\right)\right].$$

We could also use other families, such as the cubic splines that we have already discussed. Whatever choices are made, computer numerics are now involved.

13.3 The Ho and Lee Model

The basic references are three papers by Ho and Lee: [33], [34], [35] and one by Pederson, Shiu and Thorlacius [60].

We refer to the notation in Chapter 4. By (3.27)

$$F_j^n(T) = \frac{P_j^n(T+1)}{P_j^n(1)} \tag{13.38}$$

is a 1-forward price of a T-zero when in state (n, j). This motivates the following assumptions:

13.3 The Ho and Lee Model

Assumption 1

There exist (perturbation) functions $h(T) > 1 > h^*(T)$ so that

$$P_{j+1}^{n+1}(T) = \frac{P_j^n(T+1)}{P_j^n(1)} h(T) \qquad (13.39)$$

$$P_j^{n+1}(T) = \frac{P_j^n(T+1)}{P_j^n(1)} h^*(T). \qquad (13.40)$$

Note

$$\frac{1}{P_j^n(1)} = R(n,j) = 1 + r(n,j)$$

$$h(0) = h^*(0) = 1$$

$$\pi(n,j) = \frac{P_j^n(T+1)R(n,j) - P_j^{n+1}(T)}{P_{j+1}^{n+1}(T) - P_j^{n+1}(T)}$$

$$= \frac{1 - h^*(T)}{h(T) - h^*(T)},$$

and for all (n, j, T)

$$\pi(n,j)h(T) + (1 - \pi(n,j))h^*(T) = 1. \qquad (13.41)$$

Assumption 2

Bond prices are path-independent. This means that the Ho and Lee Model has a recombining tree model. This means that moving $(n,j) \to (n+1, j+1) \to (n+2, j+1)$ leads to the same value as $(n,j) \to (n+1, j) \to (n+2, j+1)$. This is equivalent to

$$\frac{h(T+1)h^*(T)}{h(1)} = \frac{h^*(T+1)h(T)}{h^*(1)} \qquad (13.42)$$

for all T.

Assumption 3

$\pi(n, j) = \pi$ is independent of (n, j).

Lemma 13.3.

$$h(T) = \frac{1}{\pi + (1-\pi)\delta^T} \qquad (13.43)$$

$$h^*(T) = \frac{\delta^T}{\pi + (1-\pi)\delta^T} \qquad (13.44)$$

Proof. Set $y(T) = h(T)^{-1}$ and show that

$$y(T+1) = \gamma + \delta y(T) \qquad (13.45)$$

where

$$\gamma = \frac{\pi(h(1)-1)}{(1-\pi)h(1)} = (1-\delta)\pi \qquad (13.46)$$

$$\delta = \frac{1-\pi h(1)}{(1-\pi)h(1)} \qquad (13.47)$$

and (13.45) has solution

$$y(T) = A + B\delta^T \qquad (13.48)$$

where

$$A = \gamma + \delta A$$
$$1 = A + B.$$

Show that $A = \pi$ and $B = 1 - \pi$ to complete the proof. $\square$

Remark 13.4. The Ho and Lee model is a two-parameter model and is completely specified by values for $P(0,n)$ for various n, and the values of π and $\delta < 1$.

We shall now write $P(n)$ for $P(0,n)$.

Lemma 13.5.

$$P_j^n(1) = \frac{P(n+1)}{P(n)} h(n) \delta^{n-j} \qquad (13.49)$$

$$P_j^n(T) = \frac{P(n+T)}{P(n)} \frac{h(T)h(T+1) \cdots h(T+n-1)}{h(0)h(1) \cdots h(n-1)} \delta^{(n-j)T} \qquad (13.50)$$

Proof. An exercise for the reader. $\square$

Remark 13.6. We require that $0 < \delta < 1$. This will imply that $h(T) > 1$ for all T (and $h^*(T) < 1$ for all T).

13.3 The Ho and Lee Model

One of the problems with the Ho and Lee model is this: Unless π and δ are chosen carefully some of the $r(n,j)$ can be negative. In fact an example was given by Ritchken and Boenawan [64]. If $P(1) = 0.9091$, $P(2) = 0.8417$, $\pi = 0.4$ and $\delta = 0.8$, then $r(1,1) < 0$. Ritchken and Boenawan also give conditions to ensure that $r(n,j) \geq 0$ for all (n,j). In fact $r(n,j) \geq 0$ if and only if $P_j^n(1) \leq 1$. This holds if $P(n+1)h(n) \leq P(n)$ for all n. This can be achieved by selecting $\delta \geq \delta_1$ where

$$\delta_1 = \max_{1 \leq n \leq T-1} \left[\frac{\frac{P(n+1)}{P(n)} - \pi}{1 - \pi} \right]^{\frac{1}{n}}. \tag{13.51}$$

As $P(n+1) < P(n)$ for all n, $\delta_1 < 1$, we can choose $\delta_1 \leq \delta < 1$.

We can now select the (π, δ) to satisfy the Ritchken and Boenawan condition and to minimize

$$\sum_{i=1}^{M} w_i \left[\tilde{C}_i - C_i(\pi, \delta) \right]^2$$

where $\tilde{C}_i$ for $i = 1, 2, \ldots, M$ are market prices of M interest rate derivatives whose Ho and Lee values for parameters (π, δ) are $C_i(\pi, \delta)$ for $i = 1, 2, \ldots, M$. Of course, $w_1, w_2, \ldots, w_M$ are some weights.

This shows one possible way to overcome one of the disadvantages of the Ho and Lee model—negative interest rates. Other weaknesses are: bond yields all have the same volatility; there is no mean reversion in interest rates. These are disadvantages when the market data do not have these properties.

Pricing Contingent Claims with Ho and Lee

There are two approaches for European claims:

First approach

Given $V(N, j)$ for $j = 0, 1, 2, \ldots, N$,

$$V(n, j) = P_j^n(1) \left[\pi V(n+1, j+1) + (1-\pi) V(n+1, j) \right] \tag{13.52}$$

using results from Lemma 13.5. This approach can readily be adapted to American claims, and so forth.

Second approach

Evaluate

$$V(0,0) = \sum_{j=0}^{N} \lambda(N, j) V(N, j) \tag{13.53}$$

using Arrow-Debreu prices (or state prices). These can be calculated as follows:

13 Interest Rate Models

Lemma 13.7.

$$\lambda(n,j) = \frac{P(n)\pi^j(1-\pi)^{n-j}\psi(j,n-j)}{\prod_{k=1}^{n}[\pi+(1-\pi)\delta^{n-k}]} \quad (13.54)$$

where $\psi(k,m)$ is calculated from

$$\psi(k,m) = \delta^m \psi(k-1,m) + \delta^{m-1}\psi(k,m-1) \quad (13.55)$$
$$\psi(0,m) = \delta^{\frac{m(m-1)}{2}} \quad (13.56)$$
$$\psi(k,0) = 1 \quad (13.57)$$

for $k \geq 0$, $m \geq 1$ in (13.55) and (13.56).

Proof. This is left as an exercise, but it follows from the Jamshidian forward induction formula for $\lambda(n,j)$. In fact, assume that (13.54) holds for $n = k$ and $j = 0,1,2,...,n$ and show the same holds for $n = k+1$ and $j = 0,1,2,...,n+1$. Use separate arguments for $j = 0$, $1 \leq j \leq n$ and $j = n+1$ separately. In fact the special cases are

$$\lambda(n,n) = \pi^n P(n) h(0) h(1) \ldots h(n-1)$$
$$\lambda(n,0) = (1-\pi)^n P(n) h(0) h(1) \ldots h(n-1) \delta^{\frac{n(n-1)}{2}}.$$

□

Example 13.8. Let $V(N,j) = \max\left[P_j^N(T) - K, 0\right]$. Then we can generalize the CRR formula for European calls (4.20) starting with

$$V(0,0) = \sum_{j=j*}^{N} \lambda(N,j)\left[P_j^N(T) - K\right]$$

where $j*$ is the integer part of

$$N + \frac{1}{T \ln \delta} \ln\left[\frac{1}{K} \cdot \frac{P(N+T)}{P(N)} \cdot \frac{h(T)h(T+1)\cdots h(T+N-1)}{h(0)h(1)\cdots h(N-1)}\right].$$

Example 13.9. Futures prices can also be computed. For example, with $G(N,j)$ equal to $P_j^N(T)$ we have

$$G(n,j) = \pi G(n+1, j+1) + (1-\pi) G(n+1, j).$$

This leads to

$$G_N = G(0,0) = \sum_{j=0}^{N} C_j^N \pi^j (1-\pi)^{N-j} P_j^N(T)$$
$$= \frac{P(N+T)}{P(N)} \Phi[\pi, \delta; N, N+T]$$
$$= F(0,0) \Phi[\pi, \delta; N, N+T].$$

Exercise 13.10. 1. Find Φ and show that it is an increasing function of δ.

2. Write down an expression for $h(T) - h^*(T)$ in terms of π and δ, and show that this is a decreasing function of δ. Thus smaller choices of δ imply greater volatility.

3. Show that $1 + r(n, j+1) = \delta[1 + r(n,j)]$ and conclude that this leads to the same interpretation.

13.4 The Pedersen, Shiu and Thorlacius Model

There are several generalizations of the Ho and Lee model. The most cited is due to Pedersen, Shiu and Thorlacius [60]. In our notation they assume in place of Assumption 3 of Section 13.3 that

$$\pi(n,j) = \pi(n) \tag{13.58}$$

for all $j = 0, 1, 2, \ldots, n$ and that

$$P_{i+1}^n(1) = P_i^n(1) c(n) \tag{13.59}$$

for all $i = 0, 1, 2, \ldots, n-1$, for suitable functions π and c.

Remark 13.11. In the Ho and Lee model, $\pi(n) \equiv \pi$ and $c(n) \equiv \frac{1}{\delta}$ are independent of n.

Using
$$P_j^n(T) = P_j^n(1) \left[\pi(n) P_{j+1}^{n+1}(T-1) + (1 - \pi(n)) P_j^{n+1}(T-1) \right] \tag{13.60}$$

and
$$P_j^n(1) = P_i^n(1) c(n)^{j-i} \quad \text{for } i, j = 0, 1, 2, \ldots, n, \tag{13.61}$$

we can obtain
$$P_j^n(T) = \prod_{k=n}^{T+n-1} g(k, T+n-1) P_j^k(1), \tag{13.62}$$

where

$$g(j,s) = 1 - \pi(j) + \pi(j)c(j+1)c(j+2)\ldots c(s)$$

for $j < s$, and

$$g(s,s) = 1,$$

and

$$P_i^n(1) = c(n)^i P_0^n(1)$$
$$= \frac{P(n+1)}{P(n)} c(n)^i \frac{\prod_{k=0}^{n-2} g(k, n-1)}{\prod_{k=0}^{n-1} g(k,n)}. \tag{13.63}$$

Remark 13.12. With the Ho and Lee model

$$g(j,s) = 1 - \pi + \pi\delta^{-(s-j)} = \frac{1}{h^*(s-j)}$$

with h^* as in (13.44), and (13.49) can be rederived.

By making restrictions on $\{c(1), c(2), \ldots\}$, there will not be any negative interest rates, nor very high interest rates in this model.

In fact, given $\{M(1), M(2), \ldots\}$ a sequence of real numbers all greater than 1, we can guarantee that

$$\frac{P(n+1)}{P(n)} \cdot \frac{1}{M(n)} \leq P_i^n(1) = \frac{1}{R(n,i)} \leq \frac{P(n+1)}{P(n)} \cdot M(n). \tag{13.64}$$

This can be achieved (see [60]) by requiring that

$$\frac{1}{M(n)} \leq c(n)^n \leq M(n). \tag{13.65}$$

Example 13.13. Jensen [44] indicated that the choices

$$\pi(n) = \pi \quad \text{a constant}$$

$$c(n) = \vartheta \left[\frac{P(n)}{P(n+1)}\right]^{\frac{1}{n}} + (1-\vartheta)\left[\frac{P(n+1)}{P(n)}\right]^{\frac{1}{n}},$$

where $0 < \vartheta < 1$, choices avoid negative interest rates.

The numbers $M(n)$ should be chosen so that the right end term of (13.64) is less than 1 (to avoid negative interest rates) and the left end term of (13.64) is not too small (to avoid unrealistically high interest rates).

Furthermore, in this model

$$P_i^n(T) = \frac{P(T+n)}{P(n)} \left[\prod_{k=0}^{n-1} \frac{g(k, n-1)}{g(k, T+n-1)}\right] \left[\prod_{j=n}^{T+n-1} c(j)\right]^i. \quad (13.66)$$

We also obtain the generalizations

$$P_i^{n+1}(T) = \frac{P_i^n(T+1)}{P_i^n(1)} h(n, T) \quad (13.67)$$

$$P_{i+1}^{n+1}(T) = \frac{P_i^n(T+1)}{P_i^n(1)} h^*(n, T) \quad (13.68)$$

with

$$h(n, T) = \frac{1}{g(n, T+n-1)} \quad (13.69)$$

$$h^*(n, T) = c(n+1)c(n+2)\ldots c(T+n-1)g(n, T+n-1). \quad (13.70)$$

Remark 13.14.

(a) We note that $P_i^n(T)$ does not depend on $\{\pi(k) \ ; \ \kappa \geq n\}$.

(b) The structure of this model is similar to the Ho and Lee model, but there are many more parameters $\{c(n)\}$ and $\{\pi(n)\}$.

(c) We could now proceed to find state (Arrow-Debreu) prices as in Section 13.3 and evaluate some basic contingent claims.

13.5 The Morgan and Neave Model

This model has appeared in the actuarial literature [55] and is quite tractable but is a one-parameter model. This parameter will be denoted by $u > 1$.
For $n = 0, 1, 2, \ldots$ set

$$R(n) = \frac{P(n)}{P(n+1)} \equiv \frac{P(0, n)}{P(0, n+1)} > 1 \quad (13.71)$$

and define for $j = 0, 1, 2, \ldots, n$

$$R(n, j) = R(n) \, u^{n-2j}. \quad (13.72)$$

We note that

$$R(n, j+1) = u^{-2} R(n, j) < R(n, j).$$

Morgan and Neave assume that $\pi(n,j) = \pi(n)$ are independent of j. This forces

$$\pi(n) = \frac{1}{1+u^{2n+1}} \tag{13.73}$$

to hold for all n, and

$$P_{j+1}^n(T) = u^{2T} P_j^n(T) \tag{13.74}$$

for all (n,j) and T.

We can further show that

$$P_0^n(T+1) = \frac{1}{u^n R(n)} \left[\pi(n)u^{2T} + (1-\pi(n))\right] P_0^{n+1}(T), \tag{13.75}$$

which leads to

$$P_0^n(T) = \prod_{k=n}^{T+n-1} \left[\frac{\pi(k)u^{2(T+n-1-k)} + (1-\pi(k))}{u^k R(k)}\right] \tag{13.76}$$

and then

$$P_j^n(T) = u^{2Tj} \prod_{k=n}^{T+n-1} \left[\frac{\pi(k)u^{2(T+n-1-k)} + (1-\pi(k))}{u^k R(k)}\right], \tag{13.77}$$

which follows from (13.74) and (13.76).

Remark 13.15. In order that interest rates do not become negative we require that $R(n,j) \geq 1$. But $R(n,n) = R(n)u^{-n}$ so u must be selected so that $R(n)u^{-n} \geq 1$ for $0 \leq n \leq N$, say. In the Ho and Lee model we had $R(n,j+1) = \delta R(n,j)$ (see Example 13.1), so there are some similarities, but this model does not have constant $\pi(n,j)$. However, there are some other similarities.

$$P_{j+1}^{n+1}(T) = \frac{P_j^n(T+1)}{P_j^n(1)} h(n,T) \tag{13.78}$$

with

$$h(n,T) = \frac{u^{2T}}{\pi(n)u^{2T} + (1-\pi(n))},$$

and

$$P_j^{n+1}(T) = \frac{P_j^n(T+1)}{P_j^n(1)} h^*(n,T) \tag{13.79}$$

with

$$h^*(n,T) = \frac{1}{\pi(n)u^{2T} + (1-\pi(n))},$$

clearly bear a similarity to the Ho and Lee model. There are generalizations in Hurlimann [38].

13.6 The Black, Derman and Toy Model

Another classic binomial model is the one due to Black, Derman and Toy [7]. They assume that

$$\pi(n,j) \equiv \frac{1}{2} \tag{13.80}$$

$$r(n,j) = r(n,0)\sigma(n)^j, \tag{13.81}$$

where $\sigma(n)$ for $n = 1, 2, \ldots$, are specified constants (usually greater than 1). We shall see that the BDT model, as it is called, can be calibrated to be consistent with observed values of $P(0, n)$ for various n, and that interest rates $r(n, j)$ are all positive. There are more parameters in this model than in the Ho and Lee model, so it is possible in BDT to also calibrate observed volatilities (or other quantities like cap prices). For a comprehensive comparison of these two models, consult Panjer et al. [59]. In particular, look at Chapter 7. We note that Sandmann and Sondermann [67], [68] produced independently a similar model at about the same time. We will also comment on a related model of Morgan and Neave [55], which has been discussed in the actuarial journals (see also [38]).

The First Calibration of the BDT Model

We assume that we are given the values of $P(0, n)$ and $\sigma(n)$ for $n = 1, 2, 3, \ldots, N$. We wish to compute $r(n, 0)$ for $n = 0, 1, 2, \ldots N-1$, and hence all interest rates by (13.81).

Using Arrow-Debreu prices,

$$\begin{aligned}
P(0, k+1) &= \sum_{j=0}^{k} \lambda(k, j) P_j^n(1) \\
&= \sum_{j=0}^{k} \frac{\lambda(k, j)}{1 + r(k, j)} \\
&= \sum_{j=0}^{k} \frac{\lambda(k, j)}{1 + r(k, 0)\sigma(k)^j}.
\end{aligned} \tag{13.82}$$

We can now calibrate

$$r(0,0) = \frac{1}{P(0,1)} - 1 \qquad (13.83)$$

and suppose that $r(n,0)$ and $\lambda(n,j)$ are known for $n = 0, 1, 2, \ldots, k-1$. We are then able to calculate $\lambda(k,j)$ from

$$\lambda(k,j) = \frac{1}{2}\left[\frac{\lambda(k-1,j)}{1+r(k-1,j)} + \frac{\lambda(k-1,j-1)}{1+r(k-1,j-1)}\right]. \qquad (13.84)$$

Here we use the usual conventions that $\lambda(n,j) = 0$ if $j > n$ or $j < 0$. Thus, the only unknown in (13.82) is $r(k,0)$.

Let us note that (13.82) uniquely determines $r(k,0) > 0$. In fact if we put $r(k,0) = 0$ in the right hand side of (13.82), we obtain $P(0,k)$, which we may assume is greater than $P(0, k+1)$. If we put $r(k,0) = \infty$, then the right hand side of (13.82) is 0, which we can assume is less than $P(0, k+1)$. By the intermediate value theorem, a choice of $r(k,0) > 0$ exists for which (13.82) holds. It is unique as the right hand side of (13.82) is a strictly decreasing function of $r(k,0)$ since each $\lambda(k,j) > 0$.

We then repeat the arguments with k replaced by $k+1$.

There are various ways of obtaining a solution of $f(x) = 0$ where

$$f(x) \equiv P(0, k+1) - \sum_{j=0}^{k} \frac{\lambda(k,j)}{1 + x\sigma(k)^j}. \qquad (13.85)$$

(a) Interval Bisection Method.

Let $x_1 = 0$, then $f(0) = P(0, k+1) - P(0, k) < 0$ as we have observed. Let $x_2 = 1 = 100\%$. We can suppose that $f(1) > 0$. Let $x_3 = 0.5\,(x_1 + x_2)$, and compute $f(x_3)$. If $f(x_3) < 0$ then replace x_1 by x_3, while if $f(x_3) > 0$ then replace x_2 by x_3. Then repeat for as long as you please until the x-values are close enough together (or the f values are sufficiently small). This method is good when you cannot use the Newton-Raphson method either because the derivative of f is not available, or it is too complicated to compute.

(b) Newton-Raphson Method.

The formula for the derivative of f is

$$f'(x) = \sum_{j=0}^{k} \frac{\lambda(k,j)\sigma(k)^j}{(1+x\sigma(k)^j)^2}. \qquad (13.86)$$

Let $x_0 = 0.5$, say. Then recursively compute

13.6 The Black, Derman and Toy Model

$$x_{n+1} = x_n - \frac{f(x_n)}{f'(x_n)}. \tag{13.87}$$

Very few iterations are required to find a good approximate solution.

(c) The Secant Method.

This is similar to the Newton-Raphson method, but we need two starting values, as in (a) where $f(0) < 0$ and $f(1) > 0$, and the derivative of f is replaced by the slope of the line joining $(x_1, f(x_1))$ and $(x_2, f(x_2))$. Then

$$x_3 = x_2 - \frac{x_2 - x_1}{f(x_2) - f(x_1)} f(x_2). \tag{13.88}$$

Then repeat, replacing x_1 by x_2 and x_2 by x_3.

Volatility Specification

There are two ways to specify volatilities: **absolute volatility** and **proportional volatility**. The convention in the markets is to use the latter, but you should always ask to make sure. Even though interest rates do not seem to change much, market people talk about 10–20% volatilities!!

Let us consider interest rates. In the multiperiod binomial model from state (n, i) where interest rate is currently $r(n, j)$ in one period, it could become $r(n+1, j+1)$ or $r(n+1, j)$ with probabilities q and $1-q$, say. Then the **absolute volatility** is the standard deviation

$$\text{AVol}(n, j) = \sqrt{q \left(r(n+1, j+1) - \bar{r}\right)^2 + (1-q)\left(r(n+1, j) - \bar{r}\right)^2} \tag{13.89}$$

where

$$\bar{r} = q \cdot r(n+1, j+1) + (1-q) \cdot r(n+1, j). \tag{13.90}$$

Therefore,

$$\text{AVol}(n, j) = \sqrt{q(1-q)} \cdot |r(n+1, j+1) - r(n+1, j)| \tag{13.91}$$

or, if $q = \frac{1}{2}$,

$$\text{AVol}(n, j) = \frac{1}{2} \cdot |r(n+1, j+1) - r(n+1, j)|, \tag{13.92}$$

which is half an **absolute** spread of the interest rates at $t = n+1$.

In a similar way we can define **proportional volatility** as the standard deviation of the logs of the interest rates

$$\text{PVol}(n,j) = \sqrt{q\left(\ln r(n+1,j+1) - \rho\right)^2 + (1-q)\left(\ln r(n+1,j) - \rho\right)^2} \tag{13.93}$$

where

$$\rho = q \cdot \ln r(n+1, j+1) + (1-q) \cdot \ln r(n+1, j) \tag{13.94}$$

and so

$$\begin{aligned}\text{PVol}(n,j) &= \sqrt{q(1-q)}|\ln r(n+1,j+1) - \ln r(n+1,j)| \\ &= \sqrt{q(1-q)} \cdot |\ln \sigma(n+1)|,\end{aligned} \tag{13.95}$$

which is independent of j if the up and down factors are independent of the state.

When $q = \frac{1}{2}$, we have a nice interpretation of the $\ln \sigma(n+1)$ as twice the proportional volatility of the one-period interest rate when viewed from $t = n$, which we shall write as $\sigma_r(n+1)$. So with this interpretation

$$\sigma(n) = \exp(2 \cdot \sigma_r(n)) \tag{13.96}$$

for each n.

It was a starting point in Sandmann and Sondermann [67] to write this as

$$r(n,j) = r(n, j+1) g(\sigma_r(n), q(n))$$

with

$$g(\sigma, q) = \exp\left[\frac{\sigma}{\sqrt{q(1-q)}}\right] > 1$$

where they allowed $q = q(n)$. They note that g has a minimum value when $q = \frac{1}{2}$. They assume then that $q(n)$ has the same value for all n as in a BDT model. The reader is referred to their paper for extensive discussions.

Let us finally note an approximate relationship between two volatilities. From (13.92)

$$\begin{aligned}\text{AVol}(n,j) &= \frac{1}{2} \cdot r(n+1,j) \cdot |\sigma(n+1) - 1| \\ &= \frac{1}{2} \cdot r(n+1,j) \cdot (\exp(2 \cdot \sigma_r(n)) - 1) \\ &\approx \frac{1}{2} \cdot r(n+1,j) \cdot (2 \cdot \sigma_r(n))\end{aligned}$$

$$= r(n+1,j) \cdot \sigma_r(n)$$
$$= r(n+1,j) \cdot \text{PVol}(n,j). \tag{13.97}$$

That is, the absolute volatility is equal to the proportional volatility times the current level of the rates. This means that PVol can be much higher than AVol. If interest rates are about 6% and AVol is 1%, then the PVol would be about 17%, by (13.97). It should be clear that the BDT model fits well with the market understanding of volatility.

We shall also be interested in yield volatilities.

Yields $Y(0,T)$ and $Y(n,j,T)$ are defined by

$$P(0,T) = [1 + Y(0,T)]^{-T} \tag{13.98}$$

and

$$P_j^{(n)}(T) = [1 + Y(n,j,T)]^{-T}. \tag{13.99}$$

Now specifying $P(0,n)$ is the same as specifying the yield $Y(0,n)$. In place of using $\sigma(n)$ values we can start with yield volatilities

$$\sigma_y(1), \sigma_y(2), \sigma_y(3), \ldots, \sigma_y(N-1), \tag{13.100}$$

which we now clarify. As above, this is the same as requiring that

$$Y(1,1,n-1) = Y(1,0,n-1) \cdot \exp(2 \cdot \sigma_y(n)) \tag{13.101}$$

where

$$\sigma_y(n) = \frac{1}{2} \ln \left[\frac{Y(1,1,n-1)}{Y(1,0,n-1)} \right]. \tag{13.102}$$

Let us write $v(n) = \exp(2 \cdot \sigma_y(n))$ from now on, for short.

A Second Calibration of BDT Model

Suppose we are now given the values of $P(0,n)$ and $v(n)$ for values of $n = 1, 2, 3, \ldots, N-1$. We must now solve for $r(n,0)$ and $\sigma(n)$.

Step 1.

Find $P_1^{(1)}(k)$ and $P_0^{(1)}(k)$ using the values of $P(0, k+1)$ and $v(k)$. As above

$$P(0, k+1) = \frac{0.5}{1 + r(0,0)} \left[P_0^{(1)}(k) + P_1^{(1)}(k) \right]$$

$$= \frac{0.5}{1+r(0,0)} \left[\frac{1}{(1+Y(1,0,k))^k} + \frac{1}{(1+v(k+1)Y(1,0,k))^k} \right],$$

which is an equation in one unknown $Y(1,0,k)$. This equation can be solved by one of the methods described in Section 13.6. Once obtained, we have

$$P_0^{(1)}(k) = \frac{1}{(1+Y(1,0,k))^k} \tag{13.103}$$

$$P_1^{(1)}(k) = \frac{1}{(1+v(k+1)Y(1,0,k))^k}. \tag{13.104}$$

$$\tag{13.105}$$

Step 2.

We now solve

$$P_0^{(1)}(k) = \sum_{j=0}^{k-1} \frac{\lambda^0(k,j)}{1+r(k,0)\sigma(k)^j} \tag{13.106}$$

$$P_1^{(1)}(k) = \sum_{j=1}^{k} \frac{\lambda^1(k,j)}{1+r(k,0)\sigma(k)^j}, \tag{13.107}$$

where $\lambda^0(k,j) \equiv A(1,0,k,j)$ is the value at $(1,0)$ of the time k Arrow-Debreu security that pays \$1 in state j. Likewise $\lambda^1(k,j) \equiv A(1,1,k,j)$ is the value at $(1,1)$ of the time k Arrow-Debreu security that pays \$1 in state j. The recursive formulas for $\lambda^0(k,j)$ and $\lambda^1(k,j)$ are, like (13.84),

$$\lambda^0(k+1,j) = \frac{1}{2}\left[\frac{\lambda^0(k,j)}{1+r(k,j)} + \frac{\lambda^0(k,j-1)}{1+r(k,j-1)}\right] \tag{13.108}$$

$$\lambda^1(k+1,j) = \frac{1}{2}\left[\frac{\lambda^1(k,j)}{1+r(k,j)} + \frac{\lambda^1(k,j-1)}{1+r(k,j-1)}\right]. \tag{13.109}$$

We define F and G by

$$F(x,y) = P_0^{(1)}(k) - \sum_{j=0}^{k-1} \frac{\lambda^0(k,j)}{1+xy^j} \tag{13.110}$$

$$G(x,y) = P_1^{(1)}(k) - \sum_{j=1}^{k} \frac{\lambda^1(k,j)}{1+xy^j}, \tag{13.111}$$

13.6 The Black, Derman and Toy Model

and we wish to solve

$$F(x, y) = 0 \tag{13.112}$$
$$G(x, y) = 0 \tag{13.113}$$

for $x = r(k, 0)$ and $y = \sigma(k)$.

For this the 2-dimensional Newton-Raphson method works. Let (x_n, y_n) be the nth approximation. We choose (x_{n+1}, y_{n+1}) so that

$$0 = F^n + (x_{n+1} - x_n)F_x^n + (y_{n+1} - y_n)F_y^n \tag{13.114}$$
$$0 = G^n + (x_{n+1} - x_n)G_x^n + (y_{n+1} - y_n)G_y^n \tag{13.115}$$

where

$$F^n \equiv F(x_n, y_n)$$
$$G^n \equiv G(x_n, y_n)$$
$$F_x^n \equiv \frac{\partial F}{\partial x}(x_n, y_n)$$

and so forth. The right hand side of (13.114) and (13.115) are approximations for $F(x_{n+1}, y_{n+1})$ and $G(x_{n+1}, y_{n+1})$, respectively.

The solution of (13.114) and (13.115) is

$$x_{n+1} = x_n - \frac{1}{\Delta^n}\left[G_y^n F^n - F_y^n G^n\right] \tag{13.116}$$

$$y_{n+1} = y_n - \frac{1}{\Delta^n}\left[F_x^n G^n - G_x^n F^n\right] \tag{13.117}$$

$$\Delta^n = F_x^n G_y^n - F_y^n G_x^n. \tag{13.118}$$

It can be shown that Δ^n is never 0 when $x_n \neq 0$ and $y_n \neq 0$.

Remark 13.16. We can show that for $x_n = x > 0$ and $y_n = y > 0$,

$$\Delta^n = \frac{x}{y}\sum_{i,j=1}^{k}\frac{\lambda^0(n,j)\,\lambda^1(n,i)}{(1+xy^j)^2(1+xy^i)^2}\,y^{i+j}\,(i-j) \tag{13.119}$$

$$= \frac{x}{y}\sum_{i>j}\frac{(i-j)\,y^{i+j}}{(1+xy^j)^2(1+xy^i)^2}\,\Theta_{i,j}^n \tag{13.120}$$

200 13 Interest Rate Models

with

$$\Theta_{i,j}^n \equiv \lambda^0(n,j)\lambda^1(n,i) - \lambda^0(n,i)\lambda^1(n,j) > 0 \tag{13.121}$$

for $n \geq i > j \geq 0$, which shows that $\Delta^n > 0$ for all $x, y > 0$ and $n \geq 1$.

We can further show that if $x_n > 0$ and $y_n > 1$ in (13.116) and (13.117), then $x_{n+1} > 0$ and $y_{n+1} > 1$.

Explicit formulae can be written down for F_x, F_y, G_x and G_y using (13.110) and (13.111). For example,

$$F_x(x,y) = \sum_{j=0}^{k-1} \frac{\lambda^0(k,j)y^j}{(1+xy^j)^2}. \tag{13.122}$$

A starting value $(x_0, y_0) = (0.10, 1.50)$ will usually work.

Remark 13.17. The book by Panjer et al. [57, pages 342–344] also describes how to calibrate BDT using cap prices.

Example 13.18. This uses the data from the original BDT paper, but we present the details differently.

Table 13.2. BDT example.

Maturity t	Yield $Y(0,0,t)$	Yield Volatility $\sigma_y(t)$
1	10.00	20
2	11.00	19
3	12.00	18
4	12.50	17
5	13.00	16

From Table 13.2, $P(0,n) = 100[1+Y(0,0,n)]^{-n}$, which can be calculated for $n = 1, 2, 3, 4, 5$. Then $r(0,0) = 0.10$.

Finding r(1,1) and r(1,0).

In fact

$$r(1,1) = Y(1,1,1)$$
$$r(1,0) = Y(1,0,1)$$

so

$$r(1,1) = r(1,0) \cdot \exp(2\sigma_y(2))$$
$$= r(1,0) \cdot \exp(0.38)$$
$$= 1.462284598 \times r(1,0).$$

Then solve the system

$$P(0,2) = \frac{50}{[1+r(0,0)][1+r(1,1)]} + \frac{50}{(1+r(0,0))(1+r(1,0))}$$

$$P(0,2) = [1+0.11]^{-2} = 0.811622433$$

$$0.8116 = \frac{50}{1.1(1+1.4622r(1,0))} + \frac{50}{1.1(1+r(1,0))}$$

which gives $r(1,0) = 0.0979$. Then $r(1,1) = 1.4622$ and $r(1,0) = 0.1432$.

Finding r(2,2), r(2,1) and r(2,0).

This is trickier. For this try:

$$u(n) = Y(1,1,n)$$
$$v(n) = Y(1,0,n).$$

Then

$$u(n) = v(n) \cdot \exp(2 \cdot \sigma_y(n+1))$$

thus

$$u(2) = v(2) \cdot \exp(0.36) = 1.433329415 \cdot v(2).$$

Also

$$P_1^{(1)}(n) = \frac{100}{(1+u(n))^2}$$
$$P_0^{(1)}(n) = \frac{100}{(1+v(n))^2}.$$

Then

13 Interest Rate Models

$$P(0, n+1) = \frac{\frac{1}{2}P_1^{(1)}(n) + \frac{1}{2}P_0^{(1)}(n)}{1 + r(0,0)}$$

and

$$P(0, n+1) = \frac{100}{(1 + Y(0, 0, n+1))^{n+1}}.$$

In this case this gives the equation for $v(2)$,

$$\frac{1}{1.12^3} = \frac{1}{1.1} \left[\frac{50}{(1 + 1.4333 v(2))^2} + \frac{50}{(1 + v(2))^2} \right],$$

which is readily solved, giving $v(2) = 0.10755311$, and so $u(2) = 0.15415903$. We then have

$$P_1^{(1)}(2) = 75.0703931$$
$$P_0^{(1)}(2) = 81.5212593.$$

This procedure is quite general for finding $u(n)$ and $v(n)$ and then the values of $P_1^{(1)}(n)$ and $P_0^{(1)}(n)$ with no additional complications.

Recall that $r(n, j) = r(n, 0) \cdot \sigma(n)^j$ so we can write down two equations for $r(n, 0)$ and $\sigma(n)$.

$$\phi(r(n,0), \sigma(n)) \equiv P_1^{(1)}(n) - \sum_{k=0}^{n} \frac{A(1, 1, n, k)}{1 + r(n, 0) \cdot \sigma(n)^k} = 0$$

$$\psi(r(n,0), \sigma(n)) \equiv P_0^{(1)}(n) - \sum_{k=0}^{n} \frac{A(1, 0, n, k)}{1 + r(n, 0) \cdot \sigma(n)^k} = 0.$$

These are equivalent to one equation

$$[\phi(r(n,0), \sigma(n))]^2 + [\psi(r(n,0), \sigma(n))]^2 = 0$$

which can be solved by SOLVER in MS-EXCEL. Doing this with $n = 2$ we are solving

$$\phi(r(2,0), \sigma(2)) \equiv P_1^{(1)}(2) - \sum_{k=0}^{2} \frac{A(1, 1, 2, k)}{1 + r(2, 0) \cdot \sigma(2)^k} = 0$$

13.6 The Black, Derman and Toy Model

$$\psi(r(2,0),\sigma(2)) \equiv P_0^{(1)}(2) - \sum_{k=0}^{2} \frac{A(1,0,2,k)}{1+r(2,0)\cdot\sigma(2)^k} = 0,$$

where

$$A(1,1,2,0) = 0$$
$$A(1,1,2,1) = 0.4504$$
$$A(1,1,2,2) = 0.4504$$
$$A(1,0,2,0) = 0.4587$$
$$A(1,0,2,1) = 0.4587$$
$$A(1,0,2,2) = 0$$

as above. As SOLVER may not be able to handle this too well, one probably has to obtain the result via a 2-dimensional Newton-Raphson method, which we have described.

One can also try the approximate method due to Bjerksund and Stensland [5]:

$$\sigma(n) \approx \left[\frac{\frac{P_1^{(1)}(n-1)}{P_1^{(1)}(n)} - 1}{\frac{P_0^{(1)}(n-1)}{P_0^{(1)}(n)} - 1} \right]$$

with

$$r(n,0) \approx \left[\frac{1+\sigma(n)}{2}\right]^{-n} \left[\frac{P(0,n)}{P(0,n+1)} - 1\right].$$

Applying this with $n = 2$ and using

$$P_1^{(1)}(1) = \frac{100}{(1+r(1,1))} = 90.0901$$
$$P_0^{(1)}(1) = \frac{100}{(1+r(1,0))} = 91.7431$$
$$P_1^{(1)}(2) = 75.0703931$$
$$P_0^{(1)}(2) = 81.5212593$$

gives

$$\sigma(2) \approx \frac{\left[\dfrac{P_1^{(1)}(1)}{P_1^{(1)}(2)} - 1\right]}{\left[\dfrac{P_0^{(1)}(1)}{P_0^{(1)}(2)} - 1\right]} = \frac{\left[\dfrac{90.0901}{75.0703931} - 1\right]}{\left[\dfrac{91.7431}{81.5212593} - 1\right]}$$

$$= \frac{0.200074973}{0.12438865} = 1.595638617,$$

and

$$r(2,0) \approx \left[\frac{1+\sigma(2)}{2}\right]^{-2} \left[\frac{P(0,2)}{P(0,3)} - 1\right]$$

$$= 0.593706136 \times \left[\frac{0.811622433}{0.711780247} - 1\right] = 0.083279801.$$

Then

$$r(2,0) \approx 0.08328$$
$$r(2,1) \approx 0.13288$$
$$r(2,2) \approx 0.21204,$$

which is reasonably close to the BDT values, and could be used as a starting value for a more accurate algorithm like the Newton-Raphson method.

Applying SOLVER as described above gives $r(2,0) = 0.09700444$ and $\sigma(2) = 1.60060528$. This yields

$$r(2,0) \approx 0.09700$$
$$r(2,1) \approx 0.15526$$
$$r(2,2) \approx 0.24852.$$

It is interesting to note that if one repeats the calculations, one ends up with $P(0,3) = 71.1772$, and $P_1^{(1)}(2) = 75.0692$ and $P_0^{(1)}(2) = 81.52078$. Therefore, one still gets a good agreement for yields and volatilities. The Bjerksund and Stensland approximation does not do so well. It seems therefore that solving for $r(n,0)$ and $\sigma(n)$ can be rather delicate. Therefore, as noted, the values obtained could be used as starting values for a Newton-Raphson method.

13.7 Defaultable Bonds

We present here a simple model based on the ideas of Merton [54]. Let $V(n,j)$ represent the value of a corporation at time $t = n$ in state j. At $t = 0$ it is funded from equity $S(0,0)$ and debt $B(0,0)$ as

$$V(0,0) = S(0,0) + B(0,0). \tag{13.123}$$

The debt becomes due at $t = N$. This means that B must be repaid by the firm at $t = N$, with no repayments before this date. This is fine if $V(N,j) \geq B$, but the debt-holders will only receive $V(N,j)$ if $V(N,j) < B$. The debt-holders therefore receive

$$\tilde{B}(N,j) = \min\left[V(N,j), B\right] \tag{13.124}$$

at $t = N$. The present value of $\tilde{B}(N, \cdot)$ is

$$B(0,0) = B\tilde{P}(0,N). \tag{13.125}$$

It is not hard to show that $\tilde{P}(0,N) < P(0,N)$ whenever there is a possibility that $V(N,j) < B$ in some state j. In that case the corporation raised a loan with defaultable debt.

At expiration of the debt $t = N$ the surplus $S(N,j) = \max[V(N,j) - B, 0]$ belongs to the shareholders, and $S(0,0)$ is the present value of this shareholder value. We came across this concept in Chapter 4 with Remark 4.15 about compound options.

Given a binomial tree with values $\{V(n,j), \pi(n,j)\}$ we can now find the value of corresponding defaultable zero-coupon bonds.

13.8 Exercises

Exercise 13.19. Compute b_0, c_0 and d_0 in equation (13.18) and discuss the graph of $m(t)$ that results.

Exercise 13.20. Solve the optimization problem in equation (13.6) using equal weights.

Exercise 13.21. Check the values of $\kappa(1S)$, $\kappa(2S)$ and $\kappa(3S)$ in equation (13.16).

Exercise 13.22. Check the values of $P(0,t)$ for $t = 1(0.5)5$ in equation (13.18). [That is, we let t take values from 1 to 5 in steps of 0.5.]

Exercise 13.23. Interpret formula (13.30).

Exercise 13.24. Verify the values for x_1 and x_2 in equation (13.33).

Exercise 13.25. Give a full proof of Lemma 13.3. It is also possible to write down and solve a recurrence relation for $\frac{h^*(T)}{h(T)}$.

Exercise 13.26. Give a full proof of Lemma 13.5. We note that (13.49) is a special case of (13.50). As a hint assume that (13.50) holds for $n = k$ and for all $T \geq 1$, and show that result holds for $n = k+1$ and all $T \geq 1$ using

$$P_j^{k+1}(T) = \frac{P_j^k(T+1)}{P_j^k(1)} h(T)\delta^T .$$

Exercise 13.27. Verify the Ritchken and Boenawan condition (13.5).

Exercise 13.28. Verify (13.51).

Exercise 13.29. Give a full proof for Lemma 13.7. Show one computes the values of $f(k, m)$ using the "boundary" values.

Exercise 13.30. Study Example 13.8.

Exercise 13.31. Study Example 13.9.

Exercise 13.32. Verify formula (13.62). As a hint, perform a mathematical induction argument with respect to T. In other words assume the formula holds for T and all (n, j) and show that the formula is again true for T replaced by $T+1$ and all (n, j). The case $T = 1$ is dealt with separately.

Exercise 13.33. Verify the formulae (13.66)–(13.70).

Exercise 13.34. Find formulae for $\lambda(n, j)$ in the Pedersen, Shiu and Thoraclius model, and find expressions for European call prices, European put prices and futures prices analogous to those obtain above in Section 13.3 for the Ho and Lee model.

Exercise 13.35. Verify the claims of Jensen's Example 13.13.

Exercise 13.36. Verify the claims of the Morgan and Neave model. To show that $\pi(n)$ has the form in (13.73) start by finding $\pi(1)$ and $\pi(2)$. Formula (13.74) can be shown by doing mathematical induction with respect to T and using the equation

$$\begin{aligned}P_j^n(T+1) &= P_j^n(1) \left[\pi(n) P_{j+1}^{n+1}(T) + (1 - \pi(n)) P_j^{n+1}(T) \right] \\ &= P_j^n(1) \left[\pi(n) u^{2T} + (1 - \pi(n)) \right] P_j^{n+1}(T).\end{aligned}$$

Exercise 13.37. Find a formula for $\lambda(n,j)$ in the Morgan and Neave model. Show that

$$\lambda(n,j) = \frac{1}{R(0)R(1)\cdots R(n-1)} \cdot \frac{1}{1+u} \cdot \frac{1}{1+u^3} \cdots \frac{1}{1+u^{2n-1}} Q_j^n(u)$$

where $Q_j^n(u)$ is a polynomial in u of degree less than or equal to n^2 and with coefficients 0 or 1. Can you find $Q_j^n(u)$ explicitly?

Exercise 13.38. This is a difficult exercise. Use the Jamshidian induction formula to show that $\Delta^n \neq 0$ when x_n and y_n are nonzero. [As a hint, study Remark 13.16. The key equation there (13.121) can be proved by mathematical induction together with (13.108) and (13.109).]

Exercise 13.39. Show that the final claim in Remark 13.16 holds.

Exercise 13.40 (Discount functions).

You are given the following annual interest rate data (obtained from Bloomberg for 19 August 2002):

$$\begin{array}{r}\text{overnight } 4.75\% \\ \text{30 day } 4.90\% \\ \text{60 } 4.95\% \\ \text{90 } 4.98\% \\ \text{120 } 5.00\% \\ \text{150 } 5.00\% \\ \text{180 } 5.02\%\end{array}$$

and the swap rates

$$\begin{array}{r}\text{1Q } 5.100\% \\ \text{2Q } 5.357\% \\ \text{3Q } 5.525\% \\ \text{4S } 5.695\% \\ \text{5S } 5.790\% \\ \text{7S } 5.965\% \\ \text{10S } 6.125\% \\ \text{15S } 6.225\% \\ \text{20S } 6.280\%\end{array}$$

1. Compute estimates for $P(0,t)$ for $t \leq 20$.

2. With this discount curve, is the following transaction profitable as of now (19 August 2002). You borrow (receive) $100,000 on 19 October 2003 and repay $110,000 on 19 June 2005.

3. With the values for $P(0,1)$, $P(0,3)$ and $P(0,5)$ from 1. find the cubic spline to fit these values to estimate $m(t) = P(0,t)$ for $t \leq 5$, and compare the values of $P(0,2), P(0,4)$ from this spline with the values obtained in 1.

14
Real Options

Since the early 1990s this area has received much attention as a new approach to capital investment decisions by firms. A standard book on this subject is Dixit and Pindyck [25]. Since then many books and articles devoted to real options have been published. A more recent book by Trigeorgis [75] gives a good review of issues and techniques. The reviews of these two books are given by Schwartz [69] and Sick [72], respectively. They provide interesting discussions about real options by two exponents of the theory.

Here are some quotes from the second book:

> It is now widely recognized that traditional discounted cash flow (DCF) approaches to appraisal of capital-investment projects, such as the standard net-present-value (NPV) rule, cannot properly capture management's flexibility to adapt and revise later decisions in response to unexpected developments [page 1].

And again

> An options approach to capital budgeting has the potential to conceptualize and quantify the value of options from active management and strategic interactions....Many of these real options (e.g., to defer, contract, shut down, or abandon a capital investment) occur naturally; others may be planned and built in at extra cost from the outset (e.g., to expand capacity or build growth options, to default when investment is staged sequentially, or to switch between alternative inputs or outputs) [page 4].

Further references for our approach of valuing options may be found in Musiela and Zariphopoulou [56], and Henderson [32], Davis [22], Smith and Nau [73], Elliott and van der Hoek [28], [27].

14.1 Examples

We shall now discuss a number of examples from the book by Trigeorgis. We have cited where they are discussed in greater detail, but we have used notation consistent with this book.

Example 14.1. [75, page 5] Suppose a company wishes to value an opportunity to invest in a project (such as a research and development effort to discover a new drug). One year later the project will generate expected cash flows of $180 million under good conditions ($V(1,1) = V(1,\uparrow) = 180$, say), or $60 million under bad conditions ($V(1,0) = V(1,\downarrow) = 60$). There is equal probability of each outcome. The project's cash flows have an expected rate of return (or risk adjusted discount rate) of $k = 20\%$, while the risk-free interest rate is $r = 8\%$. Then we can compute

$$V(0) = \frac{0.5 \times V(1,\uparrow) + 0.5 \times V(1,\downarrow)}{1+k} = \frac{0.5 \times 180 + 0.5 \times 60}{1+0.20} = 100 \quad (14.1)$$

in millions of dollars. Here we are using equation (2.27) of Chapter 2. We can now determine the value of π using

$$V(0) = \frac{\pi \times V(1,\uparrow) + (1-\pi) \times V(1,\downarrow)}{1+r},$$

that is

$$100 = \frac{\pi \times 180 + (1-\pi) \times 60}{1+0.08},$$

giving $\pi = 0.4$.

Suppose the government of the day, wishing to support the project, offers a guarantee (or insurance policy) to buy the entire output for $180 million if the bad conditions occur. The government's guarantee is like a put option giving the company the right to sell the project's value to the government for a guaranteed amount of $180 million. What is the value of this guarantee (put option)?

The traditional DCF (discounted cash flow) method would value the project without guarantee as above at $100 million, and for the value of the project with guarantee $V^*(0)$ as

$$V^*(0) = \frac{0.5 \times 180 + 0.5 \times (60+120)}{1+0.20} = 150$$

in million of dollars. So the value of the put option provided by the guarantee would be estimated as (an abandonment put) AP:

$$AP = V^*(0) - V(0) = 150 - 100 = 50$$

million of dollars. This traditional valuation assumes that the payoff of the put option (the guarantee) has the same risk and can be discounted at the same rate as that for the naked project (without the guarantee), that is

$$AP = \frac{0.5 \times 0 + 0.5 \times 120}{1 + 0.20} = 50$$

as above. This traditional DCF calculation, however, is completely wrong, since the flexibility to abandon the project for a guaranteed price would alter the project's risk and its discount rate. The correct value of V^* is in fact

$$V^*(0) = \frac{0.5 \times 180 + 0.5 \times (60 + 120)}{1 + 0.08} = 166.67, \tag{14.2}$$

and hence, the correct value of AP is

$$AP = V^*(0) - V(0) = 166.67 - 100 = 66.67. \tag{14.3}$$

This can also be found using risk-neutral probabilities:

$$AP = \frac{0.4 \times 0 + 0.6 \times 120}{1 + 0.08} = 66.67. \tag{14.4}$$

Example 14.2 (Option to defer investment). Instead of investing $I(0)$ now in a project with present value of future cash flows $V(0)$ (for which $NPV = V(0) - I(0)$ could be negative), a company could have an option to invest $I(1)$ at $t = 1$. The company would make only the investment at $t = 1$ provided $V(1) > I(1)$ and not proceed otherwise. Thus the profit at $t = 1$ will be $E(1) = (V(1) - I(1))^+$, the payoff is like a call option with strike price $I(1)$. Here, $V(1)$ represents the value at $t = 1$ of the future cash flows of the project. The net present value of the project with the option is $\max[V(0) - I(0), E(0)] \geq 0$.

In fact a company may be involved in a large development which must go through various stages, and the option to defer could apply to any stage of operations, where each stage may involve further investments.

Example 14.3 (Option to default on planned cost). As suggested above projects do not involve a single up-front payment; rather investments are made in stages. Investments could be deferred, or the project could be abandoned if conditions are not favorable.

Suppose that there will be installments in a project, with $I(0)$ invested at $t = 0$ and $I(1)$ invested at $t = 1$. Then management will pay the installment at $t = 1$, provided the project value $V(1)$ at $t = 1$ exceeds this installment, otherwise the installment will not be paid and the project defaults. The value of the project at $t = 1$ is then $E(1) = (V(1) - I(1))^+$, and then the net present value of the project is $E(0) - I(0)$ where $I(0)$ is the outlay at $t = 0$. If abandonment did not occur at $t = 1$ it could occur at future times in the life of the project. Having this option to default can make an otherwise nonviable project into a viable one. Note that

$$E(0) - I(0) > PV_0(V(1) - I(1)) - I(0)$$
$$= V(0) - PV_0(I(1)) - I(0).$$

Example 14.4 (Option to expand or the growth option). If conditions are favorable a company may at time $t = 1$, say, or at some other time, invest an amount I_E and expand operations by $\alpha\%$. The value of this position at $t = 1$ is then

$$E(1) = \max\left[V(1)(1 + \frac{\alpha}{100}) - I_E, V(1)\right]$$
$$= V(1) + \left(\frac{\alpha}{100}V(1) - I_E\right)^+.$$
(14.5)

The present value of this project is now $E(0) - I(0) > V(0) - I(0)$. Again, this added flexibility may make a nonviable project into a viable one.

Example 14.5 (Option to contract). If market conditions become weaker, a company may wish to have the option to scale down operations. Let us suppose that decisions are again made at $t = 1$. Either an outlay of $I(1)$ is made for the project whose value at $t = 1$ is $V(1)$, or a reduced amount $J(1)$ is invested, (saving $I^*(1) = I(1) - J(1)$), and the project is scaled down by $\rho\%$. The value $E(1)$ at $t = 1$ with these choices, is then

$$E(1) = \max\left[V(1) - I(1), (1 - \frac{\rho}{100})V(1) - J(1)\right]$$
$$= V(1) - I(1) + \left(I^*(1) - \frac{\rho}{100}V(1)\right)^+.$$
(14.6)

Again $E(0) - I(0) > V(0) - PV_0(I(1)) - I(0)$, so the added flexibility adds value.

Example 14.6 (Option to temporarily shut down operations).
Suppose that a project in a previous example had a 30% cash payout, so the cash revenues C in a particular year amount to 30% of the project's value.

14.1 Examples

Thus $C(1) = 0.30V(1)$. So $V(1) = C(1)$ plus the the $t = 1$ value of cash flows after $t = 1$. Each year management pays fixed costs (FC) plus variable costs (VC), but the latter only if $VC < C(1)$ (at time $t = 1$, say). If $VC \geq C(1)$ then operations are shut down for the next period. So, if $C(1) > VC$, $E(1) = V(1) - VC - FC$, and if $C(1) \leq VC$, then $E(1) = V(1) - C(1) - FC$. In summary

$$\begin{aligned} E(1) &= \max\left[V(1) - VC - FC, V(1) - C(1) - FC\right] \\ &= V(1) - FC - \min\left[VC, C(1)\right]. \end{aligned} \quad (14.7)$$

Therefore, the net present value of the project is $E(0) - I(0)$. As $E(1) > V(1) - VC - FC$ with this option, we have a greater net present value.

Example 14.7 (Option to abandon for salvage value). At any time a company has the choice to continue with a project or abandon the company with some salvage value A. The salvage value could be time- and state-dependent. It is possible also to regard the salvage value as the value of an alternative project, in which case we talk about abandoning the project and switching to another project.

Assuming a decision is made at time $t = 1$ say, then

$$E(1) = \max\left[V(1), A(1)\right] = V(1) + [A(1) - V(1)]^+$$

and so the net present value of the project is $E(0) - I(0)$ if there is a cost of $I(0)$ at $t = 0$, as usual. As $E(1) > V(1)$, the abandonment option has added value to the project. (Usually this is a perpetual option. This raises an interesting question: How do we value a perpetual option in a binomial model?)

Remark 14.8. We have just given some illustrations of the ideas behind real option valuation. Of course, these possibilities and others can be used in various combinations.

We have expressed the various options so that the underlying asset is V, the value of the project without options. If V were the price process of a tradeable asset then we could price these real options like any other financial option, perhaps using a binomial model for V. However, this is usually not the case. How to proceed is the subject of ongoing research, and you will see articles with titles like "The Valuation of Options on Non-Traded Assets" devoted to this topic. In a binomial framework, one assumption that is sometimes used is the following: Assume that there is some tradeable asset S whose value $S(0)$ is known, and for which the values $S(1,\uparrow) \neq S(1,\downarrow)$ are known. Here $(1,\uparrow)$ and $(1,\downarrow)$ are states where $V(1)$ takes values $V(1,\uparrow)$ and $V(1,\downarrow)$. Then the pricing works as if V were a tradeable asset. This is because $V(1) = x(1+r) + yS(1)$

implies that $a(1+r)+bV(1)$ can be synthesized from $(a+bx)(1+r)+(by)S(1)$. Trigeorgis calls S a **twin security**, but we have not required S and V to be perfectly correlated ($V(1) = kS(1)$). Without the concept of twin securities, researchers can use the concept of **certainty equivalence** from the actuarial sciences or **indifference pricing** to construct prices of real options.

14.2 Options on Non-Tradeable Assets

We have assumed in our derivative pricing that the payoff can be replicated in terms of an underlying tradeable asset. There are many examples where this is not the case, and there is active research at present to deal with such situations. There are now **hedgeable** and **non-hedgeable risks**. The hedgeable risks can be priced as we have described. The non-hedgeable risks are often priced using the concept of certainty equivalence or using indifference pricing. The concept of certainty equivalence is well known in the actuarial sciences. We shall employ indifference pricing in this section.

At time $t = 1$ we consider four states, $\omega_1, \omega_2, \omega_3, \omega_4$. Suppose there is a tradeable asset (a stock) denoted by S. Suppose $S(\omega_1) = S(\omega_2) = S(1,\uparrow)$ and $S(\omega_3) = S(\omega_4) = S(1,\downarrow)$. We shall also assume there is cash which we can invest under interest.

Starting with wealth x at $t = 0$ we can invest in stock and cash so that

$$x = H_0 + H_1 S(0). \tag{14.8}$$

At time $t = 1$, we shall have

$$X(1) = H_0 R + H_1 S(1) = xR + H_1(S(1) - RS(0)). \tag{14.9}$$

How should an investor choose H_1 (and hence H_0)?

An investor may have a utility function U and select risky investments using the **expected utility** criterion as discussed in the book by von Neumann and Morgenstern [77]. According to this criterion, an investment X is preferred to Y if and only if $\mathbf{E}[U(X)] > \mathbf{E}[U(Y)]$. Typically U will be an increasing and concave function. The latter condition is **equivalent** to the condition that an investor will always prefer $\mathbf{E}[X]$ to X for any risk X. We said "may have a utility function" because all we really require is that our investor have a way of making choices over uncertain outcomes; utility functions are useful in this respect, but not the only way to model choices.

An investor will then choose H_1 to maximize

$$\mathbf{E}[U(xR + H_1(S(1) - RS(0)))]. \tag{14.10}$$

14.2 Options on Non-Tradeable Assets

We shall determine this choice, called H_1^*, and $V(x)$ the resulting maximum expected utility.

We give (real world) probabilities to the four states p_1, p_2, p_3, p_4. We shall also write $p(\uparrow) = p_1 + p_2$ and $p(\downarrow) = p_3 + p_4 = 1 - p(\uparrow)$. We shall also take (for simplicity)

$$U(x) = -\exp(-\gamma x) \tag{14.11}$$

for some $\gamma > 0$. Then U is (strictly) increasing and (strictly) concave.

The expression in (14.10) becomes

$$p(\uparrow)\left[-\exp(-\gamma(xR + H_1(S(1,\uparrow) - RS(0))))\right]$$
$$+ p(\downarrow)\left[-\exp(-\gamma(xR + H_1(S(1,\downarrow) - RS(0))))\right]$$

and setting the derivative of this expression to zero yields the unique value

$$H_1^* = -\frac{1}{\gamma}(S(1,\uparrow) - S(1,\downarrow))^{-1} \ln\left[-\frac{p(\downarrow)(S(1,\downarrow) - RS(0))}{p(\uparrow)(S(1,\uparrow) - RS(0))}\right]$$
$$= -\frac{1}{\gamma}(S(1,\uparrow) - S(1,\downarrow))^{-1} \ln\left[\frac{p(\downarrow) \cdot (1-\pi)}{p(\uparrow) \cdot \pi}\right].$$

No arbitrage assumptions require that $(S(1,\uparrow) - S(0))(S(1,\downarrow) - S(0)) < 0$, so there is no problem taking logarithms above. Let us assume that

$$S(1,\downarrow) < S(0) < S(1,\uparrow).$$

Substituting H_1^* back into (14.10) gives

$V(x)$
$$= -p(\uparrow)^\pi p(\downarrow)^{1-\pi}\left[\left(\frac{RS(0) - S(1,\downarrow)}{S(1,\uparrow) - RS(0)}\right)^{1-\pi} + \left(\frac{S(1,\uparrow) - RS(0)}{RS(0) - S(1,\downarrow)}\right)^{\pi}\right]e^{-\gamma xR}$$
$$= -\exp(-\gamma xR)\frac{p(\uparrow)^\pi p(\downarrow)^{1-\pi}}{\pi^\pi (1-\pi)^{1-\pi}}.$$

$$(14.12)$$

Note that H_1^* does not depend on initial wealth x. The number π is, as usual,

$$\pi = \frac{RS(0) - S(1,\downarrow)}{S(1,\uparrow) - S(1,\downarrow)}.$$

Now consider a general claim G with values g_1, g_2, g_3, g_4 in the four states. If $g_1 = g_2$ and $g_3 = g_4$, then G is an attainable claim, which can be replicated by cash and S. Otherwise we may suppose that

$$G(1)_i = g_i = G(Y(\omega_i)) \tag{14.13}$$

for some (possibly) non-tradeable asset Y. Now write

$$V_G(x) = \max_{H_1} \mathbf{E}[U(X(1) - G(1))], \tag{14.14}$$

where we note that $V(x) = V_0(x)$. Actuarial science defines the **indifference price** for G to be that value ν so that

$$V(x) = V_G(x + \nu). \tag{14.15}$$

What this means is this: we are indifferent to the two situations, (1) we have nothing to do with G, and (2) for a liability of $G(1)$ at time $t = 1$, we are happy to have extra wealth ν at $t = 0$. Hence the term "indifference" price. Strictly speaking we have defined the indifference **asking** price, ν_a, rather than the indifference **bid** price, ν_b, which could be defined by setting $V(x) = V_{-G}(x - \nu_b)$ in an analogous way.

We would now like to calculate the value of ν and check that it agrees with what we already know in the case that G is attainable.

We can calculate the optimal H_1^{**} in (14.14). It is calculated like the H_1^* above. In fact, now

$$H_1^{**} = -\frac{1}{\gamma}(S(1,\uparrow) - S(1,\downarrow))^{-1}$$
$$\times \ln\left[\frac{(RS(0) - S(1,\downarrow))(p_1 \exp(\gamma g_1) + p_2 \exp(\gamma g_2))}{(S(1,\uparrow) - RS(0))((p_3 \exp(\gamma g_3) + p_4 \exp(\gamma g_4))}\right]$$
$$= -\frac{1}{\gamma}(S(1,\uparrow) - S(1,\downarrow))^{-1}$$
$$\times \ln\left[\frac{(1-\pi)\cdot(p_1\exp(\gamma g_1) + p_2\exp(\gamma g_2))}{\pi\cdot((p_3\exp(\gamma g_3) + p_4\exp(\gamma g_4))}\right].$$

Substitute this into (14.15) to get

$$V_G(x) = -\exp(-\gamma x R)\mathbf{E}[\exp(-\gamma H_1^{**}(S(1) - RS(0)) + \gamma G(1)]$$
$$= -\exp(-\gamma x)\frac{(p_1\exp(\gamma g_1) + p_2\exp(\gamma g_2))^\pi}{\pi^\pi}$$

14.2 Options on Non-Tradeable Assets

$$\times \frac{(p_3 \exp(\gamma g_3) + p_4 \exp(\gamma g_4))^{1-\pi}}{(1-\pi)^{1-\pi}}.$$

Note that this agrees with (14.12) when $G = 0$. If we now apply condition (14.15), we obtain

$$\nu_a(G) = \frac{1}{\gamma R}\left[\pi \ln\left(\frac{p_1 \exp(\gamma g_1) + p_2 \exp(\gamma g_2)}{p_1 + p_2}\right)\right.$$
$$\left. + (1-\pi) \ln\left(\frac{p_3 \exp(\gamma g_3) + p_4 \exp(\gamma g_4)}{p_3 + p_4}\right)\right]$$
$$= \frac{1}{\gamma R}\Big[\pi \mathbf{E}_p[\exp(\gamma G(1))|\ S(1) = S(1,\uparrow)]$$
$$+ (1-\pi)\mathbf{E}_p[\exp(\gamma G(1))|\ S(1) = S(1,\downarrow)]\Big]. \quad (14.16)$$

Similarly we can show

$$\nu_b(G) = -\frac{1}{\gamma R}\Big[\pi \mathbf{E}_p[\exp(-\gamma G(1))|\ S(1) = S(1,\uparrow)]$$
$$+ (1-\pi)\mathbf{E}_p[\exp(-\gamma G(1))|\ S(1) = S(1,\downarrow)]\Big] \quad (14.17)$$

These formulae can be summarized even further. Let

$$q_1 = \pi \frac{p_1}{p_1 + p_2} \qquad q_2 = \pi \frac{p_2}{p_1 + p_2} \quad (14.18)$$
$$q_3 = (1-\pi)\frac{p_3}{p_3 + p_4} \qquad q_4 = (1-\pi)\frac{p_4}{p_3 + p_4}. \quad (14.19)$$

Then

$$\nu_a = \nu_a(G) = \mathbf{E}_q\left[\frac{1}{\gamma R}\ln \mathbf{E}_q[\exp(\gamma G(1)) \mid S(1)]\right]$$

an expectation conditioned on the value of $S(1)$, and

$$\nu_b = \nu_b(G) = -\mathbf{E}_q\left[\frac{1}{\gamma R}\ln \mathbf{E}_q[\exp(-\gamma G(1)) \mid S(1)]\right].$$

Note the following:
1. If $g_1 = g_2 = g(\uparrow)$ and $g_3 = g_4 = g(\downarrow)$, then

14 Real Options

$$\nu_a = \nu_b = \frac{\pi g(\uparrow) + (1-\pi)g(\downarrow)}{R} = \mathbf{E}\left[\frac{G(1)}{R}\right] \equiv \pi(G), \qquad (14.20)$$

the **risk-neutral price**, as we already know. In particular $S(0) = \mathbf{E}_q\left[\frac{S(1)}{R}\right]$.

2. If $G = G_1 + G_2$ where G_1 is attainable and G_2 is not (necessarily) attainable, then

$$\nu_a(G) = \frac{1}{R}\mathbf{E}_q[G_1] + \nu_a(G_2) \qquad (14.21)$$

$$\nu_b(G) = \frac{1}{R}\mathbf{E}_q[G_1] + \nu_b(G_2). \qquad (14.22)$$

3. A more risk-averse person has a larger value of γ. ν_a is an increasing function of γ, while ν_b is decreasing in γ. (In fact the ν_b can be obtained from ν_a by replacing γ by $-\gamma$.)

4. We have the approximation

$$\nu_a(G) \approx \frac{1}{R}\left\{\mathbf{E}_q[G(1)] + \frac{\gamma}{2}\mathbf{E}_q\left[\mathbf{Var}_q[G(1) \mid S(1)]\right]\right\} \qquad (14.23)$$

$$\nu_b(G) \approx \frac{1}{R}\left\{\mathbf{E}_q[G(1)] - \frac{\gamma}{2}\mathbf{E}_q\left[\mathbf{Var}_q[G(1) \mid S(1)]\right]\right\}. \qquad (14.24)$$

5. We have the identity

$$H_1^{**} = H_1^* + \frac{\partial \nu_a}{\partial S_0}$$

where the dependence of ν_a on $S(0)$ is through π. (See the definition of π.) If we define

$$L(1) = \nu_a + \frac{\partial \nu_a}{\partial S_0}(S(1) - RS(0)) \qquad (14.25)$$

then

$$\mathbf{E}_p[U(L(1) - G(1))] = \mathbf{E}_p[U(0)] = U(0), \qquad (14.26)$$

so we are indifferent between $L(1) - G(1)$ and 0.

Example 14.9. $S(0) = 20$, $S(1,\uparrow) = 25$, $S(1,\downarrow) = 16$. Thus $\pi = 4/9$.

Part (a).

$p_1 = p_2 = p_3 = p_4 = 0.25$, $g_1 = g_3 = 8$, $g_2 = g_4 = 6$. Then

$$\nu_a = \frac{1}{\gamma}\ln\left[\frac{1}{2}(\exp(6\gamma) + \exp(8\gamma))\right]$$

$$\nu_b = -\frac{1}{\gamma}\ln\left[\frac{1}{2}(\exp(-6\gamma) + \exp(-8\gamma))\right].$$

14.2 Options on Non-Tradeable Assets

For $\gamma = 1, 2, 3$ the values are $\nu_a = 7.4334, 7.6625, 7.7698$ and $\nu_b = 6.5662, 6.3375, 6.2302$, so the spread widens with increasing risk aversion. The approximations hold for small values of γ.

These values of ν_a and ν_b do not depend on S at all, as it is clear that the risk $G(1)$ is independent of $S(1)$. It is also easy to check that

$$\mathbf{E}_p[U(\nu_a - G(1))] = U(0)$$
$$\mathbf{E}_p[U(G(1) - \nu_b)] = U(0).$$

Also $L(1,\uparrow) = L(1,\downarrow) = \nu_a$ in this case, as $\frac{\partial \nu_a}{\partial S(0)} = 0$.

Part (b).

We now change the p probabilities, leaving all else unchanged. Suppose $p_1 = 0.375$, $p_2 = 0.125$, $p_3 = 0.2$ and $p_4 = 0.3$, so now there is some correlation between G and S. When S goes up, G is more likely to go up than down; and if S goes down, G is more likely to go down. For $\gamma = 1, 2, 3$ the answers are $\nu_a = 7.4854, 7.6904, 7.7885$ and $\nu_b = 6.7005, 6.4347, 6.2986$, respectively. The values of $\frac{\partial \nu_a}{\partial S(0)}$ are $0.05421, 0.03376, 0.02317$, respectively. Note the dependence on risk aversion. In the case $\gamma = 1$, $L(1,\uparrow) = 7.75644$ and $L(1,\downarrow) = 7.26853$, and (14.26) can be verified easily. Similarly for $\gamma = 2, 3$.

Part (c).

Consider another change of the p probabilities, leaving all else unchanged. Suppose $p_1 = 0.375$, $p_2 = 0.125$, $p_3 = 0.3$ and $p_4 = 0.2$, so now there is less correlation between G and S. For $\gamma = 1, 2, 3$ the answers respectively are $\nu_a = 7.46560, 7.7989, 7.8632$ and $\nu_b = 6.8711, 6.5431, 6.3733$. The respective values of $\frac{\partial \nu_a}{\partial S(0)}$ are $0.02010, 0.01206, 0.00823$. Note the dependence on risk aversion. In the case $\gamma = 1$, $L(1,\uparrow) = 7.75644$ and $L(1,\downarrow) = 7.57556$, and (14.25) can be verified easily. Similarly this is the case for $\gamma = 2, 3$.

Part (d).

Consider another change of the p probabilities, leaving all else unchanged. Suppose $p_1 = 0.3125$, $p_2 = 0.375$, $p_3 = 0.3$ and $p_4 = 0.2$, so now there is negative correlation between G and S. For $\gamma = 1, 2, 3$ the respective answers are $\nu_a = 7.2995, 7.5653, 7.7014$ and $\nu_b = 6.51461, 6.3096, 6.2115$. The values of $\frac{\partial \nu_a}{\partial S(0)}$ are $-0.06901, -0.04634, -0.03221$. Note the dependence on risk aversion. In the case $\gamma = 1$, $L(1,\uparrow) = 6.95446$ and $L(1,\downarrow) = 7.57556$, and (14.25) can be verified easily. Similarly this is the case for $\gamma = 2, 3$.

For more general utility functions, explicit formulae are often not available. However properties 1. and 2. still hold, as do the following:

6. If $G_1 \leq G_2$ then $\nu(G_1) \leq \nu(G_2)$ where $\nu = \nu_a$ or ν_b.
7. For any G_1, G_2 and $0 \leq \alpha \leq 1$,

$$\nu(\alpha G_1 + (1-\alpha)G_2) \geq \alpha\nu(G_1) + (1-\alpha)\nu(G_2) \tag{14.27}$$

where $\nu = \nu_a$ or ν_b. This says that $\nu(G)$, which is, in general, nonlinear in G, is concave in G.

8. In place of property 4. we have the more general approximation,

$$\nu_b(G) \approx \frac{1}{R}[\pi G_u + (1-\pi)G_d] \tag{14.28}$$

with

$$G_u = \mathbf{E}_p[G \mid S(1) = S(1,\uparrow)] - \frac{\Gamma(x)}{2}\mathbf{var}_p[G \mid S(1) = S(1,\uparrow)] \tag{14.29}$$

$$G_d = \mathbf{E}_p[G \mid S(1) = S(1,\downarrow)] - \frac{\Gamma(x)}{2}\mathbf{var}_p[G \mid S(1) = S(1,\downarrow)] \tag{14.30}$$

and

$$\Gamma(x) = -\frac{u''(xR)}{u'(xR)} > 0 \tag{14.31}$$

and analogous approximations for $\nu_a(G)$ are the same except that $\Gamma(x)$ is replaced with $-\Gamma(x)$. The proofs are given in [27]. In fact

$$\mathbf{E}_p[G \mid S(1) = S(1,\uparrow)] = \frac{p_1 G_1 + p_2 G_2}{p_1 + p_2} \tag{14.32}$$

$$\mathbf{E}_p[G \mid S(1) = S(1,\downarrow)] = \frac{p_3 G_3 + p_4 G_4}{p_3 + p_4} \tag{14.33}$$

$$\mathbf{var}_p[G \mid S(1) = S(1,\uparrow)] = \frac{p_1 p_2}{(p_1 + p_2)^2}[G_1 - G_2]^2 \tag{14.34}$$

$$\mathbf{var}_p[G \mid S(1) = S(1,\downarrow)] = \frac{p_3 p_4}{(p_3 + p_4)^2}[G_3 - G_4]^2. \tag{14.35}$$

9. If H^+ and H^- are the payoffs of attainable claims with

$$H^- \leq G \leq H^+$$

then with $\pi(H^-)$ and $\pi(H^+)$ their risk-neutral prices

$$\pi(H^-) \leq \nu_b(G) \leq \nu_a(G) \leq \pi(H^+). \tag{14.36}$$

This shows that $\nu_a(G)$ and $\nu_b(G)$ are non-arbitrage prices. (See Appendix B.)

The ideas we have presented in this one-period scenario can be generalized by using more than one (risky) tradeable asset, but we must the resort to numerical methods or approximations in order to compute ν_a or ν_b (see [27] for details).

If we do not use any tradeable assets except cash in our analysis, then we compute $\nu_a(G)$ from

14.2 Options on Non-Tradeable Assets

$$\mathbf{E}_p\left[u(xR)\right] = \mathbf{E}_p\left[u((x + \nu_a(G))R - G(1))\right] \qquad (14.37)$$

and with (14.11), we obtain

$$\nu_a(G) = \frac{1}{\gamma R} \ln \mathbf{E}_p\left[\exp(\gamma G(1))\right] \qquad (14.38)$$

and similarly

$$\nu_b(G) = -\frac{1}{\gamma R} \ln \mathbf{E}_p\left[\exp(-\gamma G(1))\right], \qquad (14.39)$$

which is pricing by what actuaries call **certainty equivalence**. This leads to approximations

$$\nu_a(G) \approx \frac{1}{R}\left[\mathbf{E}_p[G(1)] + \frac{\gamma}{2}\mathrm{var}_p[G(1)]\right] \qquad (14.40)$$

$$\nu_b(G) \approx \frac{1}{R}\left[\mathbf{E}_p[G(1)] - \frac{\gamma}{2}\mathrm{var}_p[G(1)]\right]. \qquad (14.41)$$

Actuaries often use ν_a to price life insurance claims that are not hedgeable by financial instruments (see [30], Chapter 5). This indicates that **indifference pricing** provides a convenient bridge between this **certainty equivalence pricing** and the **risk-neutral pricing** for hedgeable claims.

One of the problems with this indifference pricing methodology is the need to specify the utility function u, on which the indifference price depends. If we use exponential utilities, and we know that $Y(0) = \nu_a(Y(1))$ and G is expressed as in (14.13), then using (14.16) with $G = Y$, we can compute γ and then use this γ to compute $\nu_a(G)$ from $G(1)$. An analogous calculation could be used if $Y(0) = \nu_b(Y(1))$. We could also work with general utilities and the approximate formula (14.28) to determine $\Gamma(x)$. Of course ν_a and ν_b are not market prices, but indifference prices considered from the point of view of an agent with utility function u.

Example 14.10. As an application of these ideas, let $Y(0)$ represent the present value of a firm with value $Y(1)$ at $t = 1$. $Y(0)$ could represent the **bid indifference price** of the (not publicly listed) firm from the point of view of the CEO of the firm: $Y(0) = \nu_b(Y(1))$. This same CEO may wish the value of the firm if it he/she an embedded option to liquidate for M the firm at $t = 1$. Then take $G(1) = \max[Y(1), M]$ and we ask for $G(0) = \nu_b(G(1))$.

Example 14.11. On the other hand, electricity is an example of a non-tradeable asset that must be bought at the ask-price of the generators. Here forward contracts are derivative contracts (exchange pool prices for fixed prices over some period of time) with zero present ask-price. These are ideas yet to be explored thoroughly.

Example 14.12. This is a binomial version of Example 14.10 when cash is the only tradeable asset. Let us assume that V is a price process of some not-necessarily traded security. For example, V could be the value of a non-listed firm, the value of some major project and so on. Clearly, V will depend on time. We also allow in particular V to represent the value that the CEO assigns to the firm or project. It is not necessarily a market price, but no doubt (and we will not be tempted to digress here), this value will also be affected by changes in the financial markets (for example, the mining stock sector index).

Let $V(n, j)$ denote the value of V at time $t = n$ in state j ($j = 0, 1, 2, \ldots, n$). From state (n, j) we can move "up" to $(n+1, j+1)$ or "down" to $(n+1, j)$. There does not have to be any up/down meaning associated with these moves. These up and down movements will occur with (real world) probabilities $\{p(n, j), 1 - p(n, j)\}$. Let us assume (for simplicity) that the CEO has exponential utility (which could be time- and state-dependent). Then (14.39) translates to

$$V(n,j) = -\frac{1}{\gamma R} \ln \left[p(n,j) \exp(-\gamma V(n+1, j+1)) \right. $$
$$\left. + (1 - p(n,j)) \exp(-\gamma V(n+1, j)) \right], \quad (14.42)$$

where $\gamma = \gamma(n, j)$ and $R = R(n, j)$ in general. We will take $R(n, j) = 1 + r_f \Delta t$, where r_f is the risk-free interest rate. If the project provides handouts (dividends), then the handouts at $t = n + 1$ must be added to the $V(n+1, \cdot)$ in the right hand side of (14.42) since the value at $t = n$ is the present value of the firm/project at $t = n + 1$ together with these handouts. If we have a model for V we still need to determine the γ and the probabilities. In the binomial models there is some hope with this parsimonious utility function. Often a value k of WACC (weighted average cost of capital) can be provided. If this is quoted continuously (as in [17, page 127], for example), then we have

$$V(n,j) \exp(k\Delta t) = p(n,j) V(n+1, j+1) + (1 - p(n,j)) V(n+1, j) \quad (14.43)$$

or

$$p(n,j) = \frac{V(n,j) \exp(k\Delta t) - V(n+1, j)}{V(n+1, j+1) - V(n+1, j)}, \quad (14.44)$$

so with $p(n, j)$ determined, we can back out the $\gamma(n, j) > 0$ from (14.43) uniquely, and infer the CEO's "local" risk-aversion parameter (as γ can be called). This can even be simplified further if we take

$$V(n,j) = V(0,0) u^j d^{n-j} \quad (14.45)$$

as in the Cox-Ross-Rubinstein (CRR) model, where $u = \exp(\sigma\sqrt{\Delta t})$ and $d = 1/u$. In that case

$$p(n,j) = \frac{\exp(k\Delta t) - d}{u - d}, \qquad (14.46)$$

which is the same for all (n,j). However, the values of $\gamma(n,j)$ are not the same for all (n,j).

To see this consider the example from [17]. Let $V(0,0) = V = 1000$, use a CRR tree with $u = 1.06184$ (as $\Delta t = 0.25$, it seems they used a value of $\sigma = 12\%$), $n = 0,1,2$. With $k = 15\%$ and $r_f = 5\%$, $p(n,j) = 0.80323$ for all (n,j), $R = 1 + 0.05 \times 0.25 = 1.0125$. We can now back out the risk-aversion factors $\gamma(0,0) = 0.0169258$, $\gamma(1,0) = 0.0179724$ and $\gamma(1,1) = 0.0159400$.

There are authors who consider the underlying asset as though it were tradeable. In Copeland and Antikarov [17] this is called the **MAD approach**. MAD is an acronym for **Market Asset Disclaimer** which refers to the assumption: "we are willing to make the assumption that the present value of the cash flows of the project without flexibility (i.e., the traditional NPV) is the best unbiased estimate value of the project were it a traded asset". This leads to the following valuation procedure:

Let
$$q(n,j) = \frac{V(n,j)R(n,j) - V(n+1,j)}{V(n+1,j+1) - V(n+1,j)}, \qquad (14.47)$$

then
$$G(n,j) = \frac{q(n,j)G(n+1,j) + (1 - q(n,j))G(n+1,j)}{R(n,j)} \qquad (14.48)$$

is the valuation formula for a claim on V, say. We now compare the approaches.

Example 14.13 (Abandonment option).

(a) the MAD approach.

We find $A(n,j)$, $n = 0,1,...,N$; $j = 0,1,..,n$ with $N = 2$, $X = 900$, so that

$$A(N,j) = \max[V(N,j), X]$$
$$Z(n,j) = \frac{1}{R}[qA(n+1,j+1) + (1-q)A(n+1,j)]$$
$$A(n,j) = \max[Z(n,j), X].$$

Here and in the next three examples $\{q, 1-q\} = \{0.58910, 0.41090\}$. This approach yields $A(0,0) = 1002.155$, and so the value of the embedded option to abandon at 900 is $\tilde{A}(0,0) = A(0,0) - V(0,0) = 2.155$. We give the intermediate values in the table.

(b) The indifference pricing (IP) approach.

We now use the recursion

$$A(N,j) = \max\left[V(N,j), X\right]$$
$$Z(n,j) = -\frac{1}{\gamma(n,j) \cdot R} \ln\left[p\, e^{-\gamma(n,j)A(n+1,j+1)} + (1-p)\, e^{-\gamma(n,j)A(n+1,j)}\right]$$
$$A(n,j) = \max\left[Z(n,j), X\right],$$

where the $\gamma(n,j)$ values are as in the example above. We now obtain $A(0,0) = 1005.066$ and $\tilde{A}(0,0) = A(0,0) - V(0,0) = 5.066$.

The CEO ascribes higher value to the embedded abandonment option in this second approach. The MAD approach is linear in payoffs, and Copeland and Antikarov also value the abandonment put separately and again obtain the value 2.155. However, if this were done with the second method, the abandonment put by itself is worth only 0.4427. It seems that the correct way to value an embedded real option is to determine the additional value that is added to a project by this option, as we have described. This idea can also used by a CEO to value another company—to decide the value that is added to his company as a result of a takeover.

Table 14.1. Abandonment option values.

		n=0	n=1	n=2
j=0	MAD	1002.16	947.07	900.00
	IP	1005.07	949.83	900.00
j=1	MAD			1061.84	1000.00
	IP			1061.84	1000.00
j=2	MAD				1127.50
	IP				1127.50

Example 14.14 (Option to contract (shrink to $\alpha\%$)).
(a) the MAD approach.
We find $Sh(n,j)$, $n = 0, 1, \ldots, N$; $j = 0, 1, \ldots, n$, with $N = 2$, $L = 450$, $\alpha = 0.50$, so that

$$Sh(N,j) = \max\left[V(N,j), \alpha V(N,j) + L\right]$$
$$Z(n,j) = \frac{1}{R}\left[qSh(n+1,j+1) + (1-q)Sh(n+1,j)\right]$$
$$Sh(n,j) = \max\left[Z(n,j), \alpha V(n,j) + L\right],$$

which yields $Sh(0,0) = 1001.08$, and so the value of the embedded option to abandon at 900 is $\tilde{Sh}(0,0) = Sh(0,0) - V(0,0) = 1.08$. We give the intermediate values in the table.

(b) the indifference pricing approach.

We now use the recursion

$$Sh(N, j) = \max\left[V(N, j), \alpha V(N, j) + L\right]$$
$$Z(n, j) = -\frac{1}{\gamma(n, j) \cdot R} \ln\left[pe^{-\gamma(n,j)Sh(n+1,j+1)} + (1-p)e^{-\gamma(n,j)Sh(n+1,j)}\right]$$
$$Sh(n, j) = \max\left[Z(n, j), \alpha V(n, j) + L\right],$$

where the $\gamma(n, j)$ values are again as in the example above. We now obtain $Sh(0, 0) = 1002.62$ and $\tilde{Sh}(0, 0) = Sh(0, 0) - V(0, 0) = 2.62$.

The CEO ascribes again a higher value to the embedded contraction option in this second approach.

Table 14.2. Option to contract values.

		n=0	n=1	n=2
j=0	MAD	1001.08	944.42	893.46
	IP	1002.62	945.88	893.46
j=1	MAD			1061.84	1000.00
	IP			1061.84	1000.00
j=2	MAD				1127.50
	IP				1127.50

Example 14.15 (Option to Expand (by $\beta\%$)).
(a) the MAD approach.

We find $E(n, j)$, $n = 0, 1, \ldots, N$; $j = 0, 1, \ldots, n$, with $N = 2$, $M = 100$, $\beta = 0.10$ so that

$$E(N, j) = \max\left[V(N, j), (1+\beta)V(N, j) - M\right]$$
$$Z(n, j) = \frac{1}{R}\left[qE(n+1, j+1) + (1-q)E(n+1, j)\right]$$
$$E(n, j) = \max\left[Z(n, j), (1+\beta)V(n, j) - M\right],$$

which yields $E(0, 0) = 1004.32$, and so the value of the embedded expansion option is $\tilde{E}(0, 0) = E(0, 0) - V(0, 0) = 4.32$. We give the intermediate values in the table.

(b) the indifference pricing approach.

We now use the recursion

$$E(N, j) = \max\left[V(N, j), (1+\beta)V(N, j) - M\right]$$

$$Z(n,j) = -\frac{1}{\gamma(n,j) \cdot R} \ln\left[pe^{-\gamma(n,j)E(n+1,j+1)} + (1-p)e^{-\gamma(n,j)E(n+1,j)}\right]$$
$$E(n,j) = \max\left[Z(n,j), (1+\beta)V(n,j) - M\right],$$

where the $\gamma(n,j)$ values are again as in the example above. We now obtain $E(0,0) = 1002.06$ and $\tilde{E}(0,0) = E(0,0) - V(0,0) = 2.606$.

The CEO ascribes again a lower value to the embedded expansion option in this second approach. As compared with the abandonment and contraction options, the expansion option is riskier (has a more volatile payoff) and this is penalized more heavily in the second method than in the MAD approach. An alternate interpretation comes from the observation that the expansion option is positively correlated with V (when V increases the expansion option is more valuable, and conversely) while the other two options (abandonment, contraction) are negatively correlated with V (when V decreases in value, these options become more valuable). The expansion option has a speculative character, while the abandonment and contraction options have the character of hedging risks.

Table 14.3. Option to expand values.

		n=0	n=1	n=2
j=0	MAD	1004.32	941.76	886.91
	IP	1002.62	941.76	886.91
j=1	MAD		1069.26	1000.00
	IP		1068.02	1000.00
j=2	MAD			1140.26
	IP			1140.26

Example 14.16 (Compound option (all together)).

(a) the MAD approach.

We find $C(n,j)$, $n = 0, 1, \ldots, N$; $j = 0, 1, \ldots, n$, with $N = 2$, $X = 900$, $\alpha = 0.59$, $L = 450$, $M = 100$, $\beta = 0.10$ so that

$$C(N,j) = \max\left[V(N,j), X, \alpha V(N,j) + L, (1+\beta)V(N,j) - M\right]$$
$$Z(n,j) = \frac{1}{R}\left[qC(n+1,j+1) + (1-q)C(n+1,j)\right]$$
$$C(n,j) = \max\left[Z(n,j), X, \alpha V(n,j) + L, (1+\beta)V(n,j) - M\right],$$

which yields $C(0,0) = 1006.47$, and so the value of the compound option is $\tilde{C}(0,0) = C(0,0) - V(0,0) = 6.47$. We give the intermediate values in Table 14.4.

14.2 Options on Non-Tradeable Assets

(b) the indifference pricing approach.

We now use the recursion

$$C(N,j) = \max\left[V(N,j), X, \alpha V(N,j) + L, (1+\beta)V(N,j) - M\right]$$

$$Z(n,j) = -\frac{1}{\gamma(n,j) \cdot R} \ln\left[pe^{-\gamma(n,j)C(n+1,j+1)} + (1-p)e^{-\gamma(n,j)C(n+1,j)}\right]$$

$$C(n,j) = \max\left[Z(n,j), X, \alpha V(n,j) + L, (1+\beta)V(n,j) - M\right],$$

where the $\gamma(n,j)$ values are again as in the example above. We now obtain $C(0,0) = 1007.31$ and $\tilde{C}(0,0) = C(0,0) - V(0,0) = 7.316$.

The CEO ascribes again a higher value to the embedded compound option in this second approach.

Table 14.4. Compound option values.

		n=0	n=1	n=2
j=0	MAD	1006.47	947.07	900.00
	IP	1007.31	949.83	900.00
j=1	MAD			1069.26 1000.00
	IP			1068.02 1000.00
j=2	MAD			1140.26
	IP			1140.26

Remark 14.17. (a) These CRR type examples could be easily extended to more periods. For the second approach extra values of $\gamma(n,j)$ are needed. These are easily calculated. (b) Because the approaches give different valuations, they no doubt (but not in the examples provided here) lead to different decisions to exercise the embedded real options.

Example 14.18. We consider a two-time model with four states at $t = 1$. These will be labelled $(1,j)$ for $j = 1, 2, 3, 4$ and occur with (real world) probabilities p_1, p_2, p_3, p_4, respectively. We have three assets: cash, a financial asset (S) and a non-tradeable asset (Y). We have $S(1,1) = S(1,2) = d \cdot S$ and $S(1,3) = S(1,4) = u \cdot S$ with $S(0,0) = S$. We set $q = \frac{(R-d)}{(u-d)}$. We will use exponential utility with risk-aversion parameter γ. Then $\nu_b(Y)$ can be computed as follows (see 14.16):

$$\nu(Y) = \frac{1}{R}[qY_u + (1-q)Y_d] \tag{14.49}$$

$$Y_d = -\frac{1}{\gamma} \ln\left[\frac{p_1 \exp(-\gamma Y(1,1)) + p_2 \exp(-\gamma Y(1,2))}{p_1 + p_2}\right] \tag{14.50}$$

$$Y_u = -\frac{1}{\gamma} \ln \left[\frac{p_3 \exp(-\gamma Y(1,3)) + p_4 \exp(-\gamma Y(1,4))}{p_3 + p_4} \right] \tag{14.51}$$

we can think of Y_u and Y_d as relative certainty equivalences and $Y \to \tilde{Y} = (Y_u, Y_d)$ maps Y in some sense onto the attainable claims in the financial market sense; and $\nu_b(Y) = \pi(\tilde{Y})$. Y can be regarded as a tradeable asset when $Y(1,1) = Y(1,2)$, and $Y(1,3) = Y(1,4)$, for then it can be replicated exactly by a portfolio of tradeable financial assets (cash and S). For our model we will assume that S, Y are completely specified (we know values for $S(0,0), Y(0,0) = \nu_b(Y)$ and $S(1,j), Y(1,j)$ for $j = 1,2,3,4$ with $S(1,1) = S(1,2)$, $S(1,3) = S(1,4)$ and the real world probabilities p_j for state j at $t = 1$). We could have in mind that S is some market index and Y is V the value process of a firm or project. We could calibrate this model in terms of expected returns on S, correlations between S and V and WACC values as described in Example 14.43 or by some other method as we shall see below. It is assumed that γ can be inferred.

Next we consider a contingent claim (derivative) written on Y, which we call G, and we could assume that

$$G(1,j) = g(Y(1,j)) \tag{14.52}$$

for some function g (for example, $g(x) = \max[0, x - K]$), $j = 1,2,3,4$. We then seek $\nu_b(G)$, the indifference bid price of G. In fact,

$$\nu_b(G) = \frac{1}{R}[qG_u + (1-q)G_d] \tag{14.53}$$

$$G_d = -\frac{1}{\gamma} \ln \left[\frac{p_1 \exp(-\gamma G(1,1)) + p_2 \exp(-\gamma G(1,2))}{p_1 + p_2} \right] \tag{14.54}$$

$$G_u = -\frac{1}{\gamma} \ln \left[\frac{p_3 \exp(-\gamma G(1,3)) + p_4 \exp(-\gamma G(1,4))}{p_3 + p_4} \right]. \tag{14.55}$$

We shall not continue here with this example except to point out an interesting consequence of this theory. We can write

$$G = G^{(1)} + G^{(2)} \tag{14.56}$$

where $G_u^{(1)} = G_d^{(1)}$ and $G^{(2)}$ can be replicated in the financial markets. Then $G^{(1)}$ is valued as the actuaries do it (discount certainty equivalents) and $G^{(2)}$ is valued according to financial principles (using risk-neutral expectations):

$$\nu(G) = \frac{1}{R}G_u^{(1)} + \pi(G^{(2)}). \tag{14.57}$$

In fact we can choose H_0 and H_1 so that

$$c = H_0 \cdot R + H_1 \cdot S(1,j) \qquad j = 1, 2$$
$$0 = H_0 \cdot R + H_1 \cdot S(1,j) \qquad j = 3, 4$$

where $c = G_d - G_u$. Then the choices

$$G^{(2)}(1,j) \equiv H_0 \cdot R + H_1 \cdot S(1,j)$$
$$G^{(1)}(1,j) \equiv G(1,j) - G^{(2)}(1,j)$$

satisfy these requirements with $G_u^{(1)} = G_d^{(1)} = G_u$. Equation (14.57) then gives additional structure to this pricing methodology.

Remark 14.19. We can do explicit pricing in this four-state model also using power utilities where we refer to [27]. In fact it could be argued that this is a better class of utility functions to use. With exponential utilities the pricing is independent of the wealth of the CEO. This would suggest that a small company will value a project the same way a large company does as their attitude to risk is not dependent on the level of their other assets. This is not the case with power utilities. However, the use of power utilities does not lead to simple closed-form expressions, but good approximations are available (in [27, equation (52)]). We present some of these details below.

14.3 Correlation with Tradeable Assets

We present here some ideas on the calibration of Example 14.18
Let us assume that $u > d$ (without loss of generality), and define

$$r_S(1,j) = \frac{S(1,j) - S(0,0)}{S(0,0)}$$
$$r_Y(1,j) = \frac{Y(1,j) - Y(0,0)}{Y(0,0)}$$

and suppose

$$\mathbf{E}[r_S] = \exp(\mu \Delta t) - 1$$
$$\mathbf{E}[r_Y] = \exp(k \Delta t) - 1$$

and ρ is the correlation of r_S and r_Y. Then

$$\overline{S} = \mathbf{E}[S] = S(0,0)\exp(\mu \Delta t)$$
$$\overline{Y} = \mathbf{E}[Y] = Y(0,0)\exp(k \Delta t).$$

Let

$$Y^u = \frac{p_3 Y(1,3) + p_4 Y(1,4)}{p_3 + p_4}$$
$$Y^d = \frac{p_1 Y(1,1) + p_2 Y(1,2)}{p_1 + p_2},$$

and let $\theta = Y^u - Y^d$. Given the value of θ we can compute p_1, p_2, p_3, p_4 by

$$p_1 = \frac{p_d}{Y(1,1) - Y(1,2)}\left[\overline{Y} - Y(1,2) - p_u \theta\right]$$
$$p_2 = \frac{p_d}{Y(1,2) - Y(1,1)}\left[\overline{Y} - Y(1,1) - p_u \theta\right]$$
$$p_3 = \frac{p_u}{Y(1,3) - Y(1,4)}\left[\overline{Y} - Y(1,4) + p_d \theta\right]$$
$$p_4 = \frac{p_u}{Y(1,4) - Y(1,3)}\left[\overline{Y} - Y(1,3) + p_d \theta\right],$$

where

$$p_u = p_3 + p_4 = \frac{\exp(\mu \Delta t) - d}{u - d}$$
$$p_d = p_1 + p_2 = \frac{u - \exp(\mu \Delta t)}{u - d},$$

and θ solves the equation

$$\mathbf{cov}[S(1), Y(1)] = \rho \sqrt{\mathbf{var}(S(1))}\sqrt{\mathbf{var}(Y(1))},$$

which implies

$$p_u p_d \theta^2 = \rho^2 (A + B\theta), \tag{14.58}$$

where

$$A = p_u(\overline{Y} - Y(1,3))(Y(1,4) - \overline{Y}) + p_d(\overline{Y} - Y(1,1))(Y(1,2) - \overline{Y})$$

$$B = p_u p_d(Y(1,4) + Y(1,3) - Y(1,2) - Y(1,1)),$$

14.3 Correlation with Tradeable Assets

and we assume $A > 0$. Without going into detail, it is clear that S and Y cannot be specified arbitrarily without introducing arbitrage opportunities (the return on Y cannot be arbitrarily high compared with that of S). The equation (14.58) has two solutions. We take the positive root when $\rho > 0$ and the negative root when $\rho < 0$. Of course, not all choices of ρ are consistent. The formulas for p_1, p_2, p_3 p_4 must yield positive values.

Example 14.20. We make a minor modification to the Copeland and Antikarov example already discussed. We consider a one-period model with $Y(0,0) = 1000$, $k = 0.15$, $Y(1,2) = Y(1,4) = 1061.84$, $Y(1,1) = Y(1,3) = 941.76$ as before. $S(0,0) = 100$, $S(1,1) = S(1,2) = 90.48$, $S(1,3) = S(1,4) = 110.52$, which corresponds to volatility of 20% over the time $\Delta t = 0.25$. We take $\mu = 0.08$. We then have the values $\overline{Y} = 1038.21$, $\overline{S} = 102.02$, $p_u = 0.5759$, $p_d = 0.4241$, $R = 1.0125$ (as before), $q_u = 0.5374$, $q_d = 0.4626$, $A = 2278.53$, $B = 0$. We shall price the one-period abandonment option, which pays $G(1,j) = \max[X, Y(1,j)]$ where X is the abandonment value.

case(i) $\rho = 0.40$.

Then $\theta = 38.6344$, $\gamma(0,0) = 0.02013$, $p_1 = 0.1620$, $p_2 = 0.2621$, $p_3 = 0.0347$, $p_4 = 0.5411$.

Using the MAD methodology of Copeland and Antikarov ignores the financial asset. Assuming complete markets with $X = 1000$, MAD would obtain the present value of Y with abandonment as 1023.63 for $G(0,0)$.

In our approach we do not assume the market is complete and obtain 1029.97. We note that $\rho = 0.60$ gives negative probabilities without changing the possible values that Y can take.

case (ii) $\rho = 0$.

Then $\theta = 0$, $\gamma(0,0) = 0.01693$, $p_1 = 0.0835$, $p_2 = 0.3407$, $p_3 = 0.1133$, $p_4 = 0.4626$.

Using the MAD methodology of Copeland and Antikarov gives the same results as before as it ignores the correlation between S and Y. In our approach we do not assume the market is complete and obtain 1030.63.

case (iii) $\rho = -0.40$.

Then $\theta = -38.6344$, $\gamma(0,0) = 0.02804$, $p_1 = 0.0049$, $p_2 = 0.4193$, $p_3 = 0.1919$, $p_4 = 0.3840$.

Using the MAD methodology gives the same results as before.

In our approach we obtain 1030.13. We note that $\rho = -0.55$ gives negative probabilities without changing the possible values that Y can take.

Remark 14.21. We note that the risk-aversion parameter changes with the correlation because we do not change the possible outcomes for S and Y, so different risk aversions are implied by different views on correlations. These then impact on the valuation of the project with abandonment.

We remark that the calculations above were performed with MS-EXCEL and can be extended to multiperiod examples. The number of scenarios at $n = 1$ was 4; at $n = k$ it is $(1 + k)^2$, but the computations are similar.

Generalizations

The construction we have just given extends to many periods. We choose $Y(1,2) = Y(1,4)$ and $Y(1,1) = Y(1,3)$, then

$$\theta = \rho \left[\frac{A}{p_u p_d} \right]^{\frac{1}{2}}$$

and we compute probabilities p_1, p_2, p_3, p_4 as above.

We then compute $\gamma = \gamma(0,0)$. In moving to $t = 2$ we have four similar constructions with $(S(0,0), Y(0,0))$ replaced by, for example, $(u^e S(0,0), \nu^f Y(0,0))$, with $u = \exp(\sigma_S \sqrt{\Delta t})$, $\nu = \exp(\sigma_Y \sqrt{\Delta t})$ and $e, f \in \{-1, 1\}$. The probabilities will be the same as above, but the new risk-aversion parameter will be $\nu^{-f}\gamma$ for each choice of f. We can now construct the model for (S, Y) in MS-EXCEL and price claims on it. This was done with the example for the abandonment for $X = 900$ American option.

Copeland and Antikarov obtain 1002.15 and exercise (abandonment) will only occur at $t = 2$ when Y has its smallest value. For our approach we get different values for the project with embedded option depending on the correlation ρ. These are 1004.73 ($\rho = 0.4$), 1004.93 ($\rho = 0.2$), 1005.06 ($\rho = 0$), 1005.14 ($\rho =$ -0.2), 1004.61 ($\rho =$ -0.4). The exercise occurs in the same states as in Copeland and Antikarov.

A similar set of comparisons can readily be performed for other choices of abandonment level X. When we consider abandonment at $t = 1$ or $t = 2$ with $X = 1105$, different decisions result. Copeland and Antikarov would give abandonment at $t = 1$ in each state, while in our approach (with $\rho = 0.4$) abandonment is optimal in only one of the possible states (when Y is lowest). This suggests that the different methodologies can lead to different optimal decisions by our CEO.

Remark 14.22. As remarked earlier a study of this example with power utilities will result in different pricing depending on the level of assets of the CEO, and consequent different decisions. The Copeland and Antikarov approach does not result in different valuations and decisions when risk aversions change, correlations with financial markets change and when level of wealth of the company in which a project is considered changes. The indifference approach can capture these different aspects.

14.4 Approximate Methods

For general utilities, some approximations are given in [27].

These are good approximations when the time step is not too large. Using the notation and model of Section 14.3 above:

$$\nu^b(G) \approx \frac{1}{R}[q_u G_u + q_d G_d],$$

where in place of (14.54) and (14.55)

$$G_d = G^d + \frac{1}{2}\frac{u''(\theta_d^0)}{u'(\theta_d^0)}\operatorname{var}_d(G)$$

$$G_u = G^u + \frac{1}{2}\frac{u''(\theta_u^0)}{u'(\theta_u^0)}\operatorname{var}_u(G)$$

$$G^d = \frac{p_1 G_1 + p_2 G_2}{p_1 + p_2}$$

$$G^u = \frac{p_3 G_3 + p_4 G_4}{p_3 + p_4}$$

$$q_d = \frac{S(1,1) - RS(0,0)}{S(1,1) - S(1,0)}$$

$$q_u = \frac{RS(0,0) - S(1,0)}{S(1,1) - S(1,0)} = 1 - q_d$$

and

$$\operatorname{var}_d(G) = \frac{p_1 p_2}{p_d^2}(G_1 - G_2)^2$$

$$\operatorname{var}_u(G) = \frac{p_3 p_4}{p_u^2}(G_3 - G_4)^2,$$

where (θ_d^0, θ_u^0) are the optimal solution of

$$\max\left[p_d u(\theta_d^0) + p_u u(\theta_u^0)\right]$$

$$\frac{1}{R}\left[q_d \theta_d^0 + q_u \theta_u^0\right] = x$$

The proofs of the first approximation are given in [27, equation (34)]. The second approximation is [ibid., (56)]:

$$\frac{u''(\theta_d^0)}{u'(\theta_d^0)} \approx \frac{u''(\theta_d^0)}{u'(\theta_d^0)} \approx \frac{u''(xR)}{u'(xR)} \ .$$

We will now write

$$\Gamma = -\frac{u''(xR)}{u'(xR)} > 0$$

and so

$$\nu_b(G) \approx \frac{1}{R}\left[q_d G^d + q_u G^u - \frac{1}{2}\Gamma\left[q_d\frac{p_1 p_2}{p_d^2}(G_1-G_2)^2 + q_u\frac{p_3 p_4}{p_u^2}(G_3-G_4)^2\right]\right]. \tag{14.59}$$

Let us note that when $u(x) = -\exp(-\gamma x)$, then $\Gamma = \gamma$ and when $u(x) = \frac{1}{\alpha}x^\alpha$ then $\Gamma = \frac{1-\alpha}{xR}$. When we ignore the financial asset (and we only have two states at $t = 1$) then the formula further simplifies to (as now $p_1 + p_2 = 1$):

$$\nu_b(G) \approx \frac{1}{R}\left[p_1 G_1 + p_2 G_2 - \frac{1}{2}\Gamma p_1 p_2 (G_1-G_2)^2\right]. \tag{14.60}$$

If we now require that $\nu_b(Y(1,\cdot)) = Y(0,0) = Y$ we infer that

$$\Gamma = \frac{2\left[q_d Y^d + q_u Y^u - RY\right]}{\left[q_d\frac{p_1 p_2}{p_d^2}(Y_1-Y_2)^2 + q_u\frac{p_3 p_4}{p_u^2}(Y_3-Y_4)^2\right]}, \tag{14.61}$$

where we wrote $Y_j = Y(1,j)$ for brevity.

We also note that when Y at $t = 0$ and $t = 1$ are scaled as $Y \to \beta Y$, then Γ scales as $\Gamma \to \frac{\Gamma}{\beta}$. This is quite useful in the multiperiod models as indicated in the previous section. (The p_i do not change under this scaling.)

We also note that if a CEO approximately prices a project as the (risk-free) discounted (real world) expected payoff, then the CEO is assuming something like 14.60 with $\Gamma \approx 0$. This would be the case when $\gamma \approx 0$ with exponential utilities, $\alpha \approx 1$ with power utilities, or x (wealth of CEO's company) is large (with power utilities). Using 14.59 or 14.60 we do not have to be concerned which utility our CEO was using.

We now give some results using this approximate theory which is very easy to implement in MS-EXCEL. We give results for the abandonment option (as above) and give a table of comparisons. We use $X = 900$ as in Copeland and Antikarov, we use varying correlations (ρ) and MAD indicates the results from [17], Exp are the results from section 14.3, and Approx the results obtained from using the approximate theory of this section. These are the abbreviations used in Table 14.5.

Table 14.5. Approximate option values.

ρ	MAD	Exp	Approx
-0.4	1002.15	1004.61	1004.85
-0.2	1002.15	1005.14	1004.77
0.0	1002.15	1005.06	1004.68
0.2	1002.15	1004.93	1004.60
0.4	1002.15	1004.73	1004.51

Remark 14.23. 1. As in Example 14.18 it can be shown that any payoff G can be written as $G = G^{(1)} + G^{(2)}$ with $G^{(2)}$ attainable (that is, $G^{(2)}(1,1) = G^{(2)}(1,2)$, and $G^{(2)}(1,3) = G^{(2)}(1,4)$), and $G_d^{(1)} = G_u^{(1)}$ (notation as above), and then, (like (14.57)),

$$\nu^b(G) = \frac{G_u^{(1)}}{R} + \pi(G^{(2)}).$$

In fact, this decomposition is not unique. However if we set $G^{(2)}(1,3) = G^{(2)}(1,4) = G_u$ and $G^{(2)}(1,1) = G^{(2)}(1,2) = G_d$, then $G_d^{(1)} = G_u^{(1)} = 0$, and so $\nu_b(G) = \pi(G^{(2)})$.

2. Clearly the approximate scheme works well and is very easy to implement. Obviously it can be applied to many other more interesting examples.

3. With this approximation this model is almost as easy to apply as the standard binomial model.

14.5 Exercises

Exercise 14.24. A project is modelled by a four-step (year) CRR tree with $V(0,0) = \$1000$, $u = 1.06184$, $d = 1/u$, $R = 1 + 0.05$ in each period. For simplicity you may assume the existence of a "twin security". See also pages 129-139 of Copeland and Antikarov [17].

1. The project can be abandoned at any time for $900. Determine the value of this abandonment option.

2. It is possible to shrink (contract) operations by 50% at any time by selling off equipment (say) for $450. (An alternative could be to scale down operations and sublease plant equipment for $450.) Find the value of this option.

3. At times $t = 1$ and $t = 2$ it is possible for an extra $100 to expand operations by 10%. Find the value of this option.

4. What is the value of the option to do any of the actions described in 1., 2. and 3.?

Exercise 14.25. Use the same data as in Exercise 14.24: $S(0,0) = \$90$, $r = 8\%$, $T = 1$, $N = 10$, $\Delta t = 0.10$ and $\sigma = 10\%$. Assume two dividends of $\$4$ are paid. Suppose $t = 0.4$ ($n = 4$) and $t = 0.9$ ($n = 9$) are the ex-dividend dates. Compute the European and American call prices for expiry at T and strike prices $K = 30$, $K = 100$. When should there be early exercise? Discuss.

A

The Binomial Distribution

The **complementary binomial distribution function** arises naturally in the CRR version of the multistep binomial pricing model when we price call options. This appendix provides some background for these ideas.

A.1 Bernoulli Random Variables

A random variable X is said to be a **Bernoulli random variable** if it takes one of the values 1 and 0 with probabilities p (of "success") and $1-p$ (of "failure") respectively. We write

$$Pr[X = 1] = p$$
$$Pr[X = 0] = 1 - p.$$

The expected (average) value of X is given by

$$\mathbf{E}[X] = 1 \times Pr[X = 1] + 0 \times Pr[X = 0] = p. \tag{A.1}$$

As

$$\mathbf{E}[X^2] = 1^2 \times Pr[X = 1] + 0^2 \times Pr[X = 0] = p,$$

the variance is given by

$$\mathbf{Var}[X] = \mathbf{E}[(X - \mathbf{E}[X])^2]$$
$$= (1-p)^2 \times Pr[X = 1] + (0-p)^2 \times Pr[X = 0]$$

A The Binomial Distribution

$$= (1-p)^2 \times p + p^2 \times (1-p) = p(1-p).$$

That is, in summary

$$\mathbf{Var}[X] = \mathbf{E}[X^2] - \mathbf{E}[X]^2 = p(1-p). \qquad (A.2)$$

In general

$$\mathbf{E}[g(X)] = g(1) \times Pr\,[X=1] + g(0) \times Pr\,[X=0] = pg(1) + (1-p)g(0).$$

We shall also write

$$X \sim \mathcal{B}(p)$$

to indicate that X is a Bernoulli random variable with parameter p (indicating the probability of "success").

Remark A.1. Let Y be defined by

$$Y \equiv b + (a-b)X$$

where $X \sim \mathcal{B}(p)$; then Y takes values a and b with respective probabilities p and $1-p$. We also have

$$\mathbf{E}[Y] = ap + b(1-p) \quad \text{and} \quad \mathbf{Var}[Y] = (a-b)^2 p(1-p).$$

Remark A.2. Let $X \sim \mathcal{B}(p)$ then

$$Pr\,[X \leq x] = \begin{cases} 1 & \text{if } x \geq 1 \\ 1-p & \text{if } 0 \leq x < 1 \\ 0 & \text{if } x < 0. \end{cases}$$

is the distribution function of X.

Applications

- Consider a coin toss. If heads H comes up, put $X = 1$ and if tails T comes up, put $X = 0$. Then $X \sim \mathcal{B}(0.5)$, unless the coin is biased (e.g., a two-headed coin!).

- Consider a student writing an examination, and put $X = 1$ if the student passes, and $X = 0$ if the student fails. Then $X \sim \mathcal{B}(p)$, where p is the probability that the student will pass (a success).

- Throw a die. If 5 or 6 comes up, put $X = 1$ and if 1, 2, 3 or 4 comes up put $X = 0$. Then $X \sim \mathcal{B}(\frac{1}{3})$.
- Perform a test on a piece of machinery (e.g., test whether a motor is working). Put $X = 1$ if the test is successful (the motor works) and $X = 0$ if not. Then $X \sim \mathcal{B}(p)$, where p is the probability that the motor works.

A.2 Bernoulli Trials

We say that a sequence of random variables

$$X_1, X_2, X_3, \ldots X_n, \ldots$$

is a sequence of **Bernoulli trials** if these random variables are independent and all the terms are Bernoulli random variables with the same parameter. That is, for some $0 < p < 1$

$$X_n \sim \mathcal{B}(p)$$

for all n.

Example A.3. (a) Consider a sequence of coin tossings. Put $X_n = 1$ if H came up on the nth toss, and $X_n = 0$ if T came up.
(b) Consider a class of identical (similar, at least) students. They all write examinations independently (no copying). Set $X_n = 1$ if the nth student passes, and $X_n = 0$ if the nth student fails.

A.3 Binomial Distribution

How to Count

Before discussing repeated Bernoulli trials, we shall discuss binomial coefficients. Combinatorial probabilities are related to the number of possible outcomes of a random experiment.

Suppose there are 2 ways of driving from A to B, and 3 ways of driving from B to C. The number of ways of driving from A to C via B is then 6.

This is a special case of the result that if there are p ways of performing operation (1) and q ways of performing operation (2), and the trials are independent, then there are pq ways of performing operation (1) followed by (2). This is an example of the multiplication principle.

Suppose now we have k boxes labelled with the numbers $1, 2, \ldots, k$ and $n \geq k$ distinct balls labelled $1, 2, \ldots, n$. In how may ways can we put k of the n balls into the k distinct boxes?

Consider box 1. We have n labelled balls, so box 1 can be filled in n ways.

Consider then box 2. Having filled box 1, there are now $(n-1)$ balls remaining, so box 2 can be filled in $(n-1)$ ways. Therefore, box 1 and box 2 can be filled in $n(n-1)$ ways.

There will be $(n-2)$ remaining balls to fill box 3 giving $n(n-1)(n-2)$ ways to fill the first 3 boxes. Continuing, there are $n(n-1)(n-2)\ldots(n-k+1)$ ways to fill the k boxes.

Recall for any positive integer n that $n! = n(n-1)\cdots 3 \cdot 2 \cdot 1$. Also, by convention $0! = 1$.

Therefore, the number of ways of selecting k balls from n balls, **when the order of selection is important**, (because the boxes in which the balls are placed are numbered), is

$$n(n-1)\cdots(n-k+1) = \frac{n!}{(n-k)!}.$$

We wish to determine the number of ways of selecting k balls from n balls when **the order of selection is not important**. Write C_k^n (read: "n choose k"), for this number.

Now the solution to the problem of filling k numbered boxes from the n numbered balls could be obtained by the two following operations:

1. First choose any k balls from the n.
2. Then fill the k labelled boxes with the k chosen balls.

Step 1. can be done in C_k^n ways (the number we wish to find). Step 2. can be done in $k!$ ways. Therefore, step 1. followed by step 2. can be done in $C_k^n \cdot k!$ ways. However, this must equal our first answer of $\frac{n!}{(n-k)!}$. Therefore,

$$C_k^n = \frac{n!}{k!(n-k)!}.$$

Sums of Bernoulli Random Variables

Let

$$X_1, X_2, X_3, \ldots X_n, \ldots$$

be a sequence of Bernoulli trials with $X_n \sim \mathcal{B}(p)$ for each n. Put

A.3 Binomial Distribution

$$S_n = X_1 + X_2 + X_3 + \ldots + X_n.$$

Then the value of S_n is the number of outcomes of 1 in the first n trials. It is clear that S_n could take any of the values $0, 1, 2, \ldots, n$, depending on how many successes there are. We are interested in the quantity

$$Pr\,[S_n = j],$$

the probability that we have j successes in n trials. In fact

$$Pr\,[S_n = j] = C_j^n p^j (1-p)^{n-j} \qquad (A.3)$$

as there are

$$C_j^n = \frac{n!}{j!(n-j)!} = \frac{n(n-1)(n-2)\ldots(n-j+1)}{j!}$$

ways of selecting j successes from n, and each way occurs with probability $p^j(1-p)^{n-j}$. We assumed that the trials are independent, and so we can multiply probabilities.

A random variable Z that takes the values $0, 1, 2, \ldots, n$ with probabilities given by

$$Pr\,[Z = j] = C_j^n p^j (1-p)^{n-j}$$

is said to be a binomial random variable with parameters n and p, and we shall write $Z \sim \mathcal{B}(n, p)$. So $S_n \sim \mathcal{B}(n, p)$. Many writers also use the notation

$$b(k; n, p) \equiv C_j^n p_j (1-p)^{n-j},$$

where the left is shorthand for the right (there are no subscripts, superscripts—8 characters versus 13).

We now calculate means and variances.

$$\begin{aligned}\mathbf{E}[S_n] &= \mathbf{E}[X_1] + \mathbf{E}[X_2] + \ldots + \mathbf{E}[X_n] \\ &= p + p + \cdots + p \\ &= np\end{aligned}$$

and

$$\begin{aligned}\mathbf{Var}[S_n] &= \mathbf{Var}[X_1] + \mathbf{Var}[X_2] + \ldots + \mathbf{Var}[X_n] \\ &= p(1-p) + p(1-p) + \ldots + p(1-p) \\ &= np(1-p)\end{aligned}$$

Example A.4. Here is an interesting example of how the binomial distribution can be used. It is an **electricity supply problem**. Suppose that there are $n = 10$ workers who wish to use electrical power intermittently. We are interested in estimating the total load to be expected. For a coarse approximation, imagine that at any given time each worker has a probability p of requiring one unit of power. If they work independently, the probability of exactly k workers requiring (a unit amount of) power should be $b(k; n, p)$ with $n = 10$. If, on the average, a worker uses (a unit amount of) power 12 minutes per hour, we should put $p = 0.2 = (\frac{12}{60})$. The probability of seven or more workers requiring (a unit amount of) power at the same time is then

$$b(7; 10, 0.2) + b(8; 10, 0.2) + \ldots + b(10; 10, 0.2) = 0.0008643584.$$

In other words, if the supply is adjusted to six power units (available at any time), an overload has the probability of $0.00086\ldots$ meaning that it should be expected for about one minute in 1157 ($1/1157 = 0.000864304$), that is, for about one minute in 24 hours ($24 \times 60 = 1440$ minutes).

This example showed an interest in

$$\sum_{j=7}^{10} b(j; 10, 0.2),$$

or in general

$$\sum_{j=a}^{n} b(j; n, p) = Pr\left[S_n \geq a\right]. \tag{A.4}$$

This is an expression defining the **complementary binomial distribution** function, and we write the right hand side of (A.4) as

$$\Phi(a; n, p).$$

It is sometimes useful to allow noninteger values of a in (A.4). For example, note that for $a = 5.7$, $Pr[S_n \geq 5.7] = Pr[S_n \geq 6]$. Hence

$$\Phi(5.7; n, p) = \Phi(6; n, p)$$

because $j \geq 5.7$ is the same as $j \geq 6$. If $a \geq 0$ is not an integer, the sum in (A.4) is taken from $1 + \lfloor a \rfloor$, 1 plus the integer part of a (the floor function of a). We can write for any a

$$Pr\left[S_n > a\right] = \Phi(1 + \lfloor a \rfloor; n, p). \tag{A.5}$$

For example, if $a = 5.7$ then $j > 5.7$ iff $j \geq 6 = \lfloor 5.7 \rfloor + 1 = 5 + 1$. If a is not an integer

$$Pr\left[S_n > a\right] = Pr\left[S_n \geq a\right] = \Phi\left(1 + \lfloor a \rfloor; n, p\right).$$

However, if a is an integer, then

$$Pr\left[S_n \geq a\right] = Pr\left[S_n > a\right] + b(a; n, p).$$

This means that there is a jump in the function

$$a \to \Phi\left(a; n, p\right)$$

at each integer. Also, if a is not an integer,

$$Pr\left[S_n < a\right] = Pr\left[S_n \leq a\right]$$
$$= \sum_{j=0}^{\lfloor a \rfloor} b(j; n, p)$$
$$= 1 - \Phi(a; n, p)$$
$$= \Psi(a; n, p).$$

$\Psi(a; n, p)$ is called the **binomial distribution function** with parameters n and p.

A.4 Central Limit Theorem (CLT)

Let $X_1, X_2, \ldots, X_n, \ldots$ be a sequence of independent but identically distributed random variables. We focus on the case $X_n \sim \mathcal{B}(p)$ for each $n \geq 1$. Suppose that $\mathbf{E}[X_n] = \mu$ and $\mathbf{Var}[X_n] = \sigma^2$ for each $n \geq 1$. If $X_n \sim \mathcal{B}(p)$ for each $n \geq 1$, then $\mu = p$ and $\sigma = \sqrt{p(1-p)}$, as we saw in Section A.1. Let

$$S_n = X_1 + X_2 + \ldots + X_n$$

and

$$S_n^* = \frac{S_n - \mathbf{E}[S_n]}{\sqrt{\mathbf{Var}[S_n]}}.$$

Then $\mathbf{E}[S_n^*] = 0$ and $\mathbf{Var}[S_n^*] = 1$. Write

$$G_n(x) = Pr\left[S_n^* \leq x\right],$$

which is the same as

$$G_n(x) = Pr\left[S_n \leq np + x\sqrt{np(1-p)}\right]$$

in our application. As usual, write

$$\mathcal{N}(x) = \frac{1}{\sqrt{2\pi}} \int_{-\infty}^{x} e^{-\frac{1}{2}u^2} du.$$

This is the **normal distribution function**. The values of $\mathcal{N}(x)$ are tabulated in many books (statistics texts, finance texts, and so on), but are given with greatest accuracy in books like Abramowitz and Stegun, [1] which give values to 15 decimal places. There are also analytic approximations like the following, which can be incorporated in computer programs:

$$\mathcal{N}(x) \approx \begin{cases} 1 - n(x)\left(a_1 k + a_2 k^2 + a_3 k^3\right) & \text{if } x \geq 0 \\ 1 - \mathcal{N}(-x) & \text{if } x < 0. \end{cases}$$

Here

$$k = \frac{1}{1 + \alpha x}, \quad \alpha = 0.33267$$

$$a_1 = 0.4361836, \quad a_2 = -0.1201676, \quad a_3 = 0.9372980$$

$$n(x) = \frac{1}{\sqrt{2\pi}} \exp\left(-\frac{1}{2}x^2\right).$$

This approximation is accurate to about 4 decimal places, (with a maximum error bound of 0.0002). For more accurate expressions (e.g., to 7 decimal places), see Abramowitz and Stegun's book.

The central limit theorem (CLT) says that as $n \to \infty$

$$G_n(x) \to \mathcal{N}(x)$$

for each real x. Of course this means that

$$Pr\left[S_n \leq np + x\sqrt{np(1-p)}\right] \approx \mathcal{N}(x)$$

or

$$Pr[S_n \leq x] \approx \mathcal{N}\left(\frac{x-np}{\sqrt{np(1-p)}}\right). \tag{A.6}$$

In fact the approximation

$$Pr[S_n \leq x] \approx \mathcal{N}\left(\frac{x+\frac{1}{2}-np}{\sqrt{np(1-p)}}\right) \tag{A.7}$$

is quite accurate when $\min[np, n(1-p)] > 10$ (if $p = 0.2$, this means $n > 50$).

Example A.5. Let us calculate

$$\Psi(6; 10, 0.2) = \sum_{j=0}^{6} b(j; 10, 0.2)$$

using (A.6). Now

$$\begin{aligned}
\Psi(6; 10, 0.2) &\approx \mathcal{N}\left(\frac{6 - 10 \times 0.2}{\sqrt{10 \times 0.2 \times 0.8}}\right) \\
&= \mathcal{N}\left(\frac{4}{\sqrt{1.6}}\right) \\
&= \mathcal{N}(3.16227766) \\
&\approx \mathcal{N}(3.15) + \frac{0.01227766}{0.05}[\mathcal{N}(3.20) - \mathcal{N}(3.15)] \\
&= 0.9991836477 + 0.2455532 \times 0.000129215 \\
&= 0.999215376.
\end{aligned}$$

So $Pr[S_{10} \geq 7] \approx 0.00078462$, which agrees reasonably well with 0.00086435. The choice $n = 10$ is a little small. Traditional practice says we need n to be about 30 or more.

A.5 Berry-Esséen Theorem

This theorem says something about the difference between $G_n(x)$ and $\mathcal{N}(x)$ defined above. In fact **Berry** (1941) and **Esséen** (1945) (see Chung [15]) showed that

$$\sup_x |G_n(x) - \mathcal{N}(x)| \leq C \frac{\mathbf{E}(|X_1 - \mu|^3)}{\sigma^3 \sqrt{n}} \tag{A.8}$$

for all n. This means that the difference magnitude $|G_n(x) - \mathcal{N}(x)|$ is less than the right hand side in (A.8) for all x. In fact Berry and Esséen gave $C = \frac{33}{4}$, but a more precise argument gives $C \approx 0.7975$, and it can be shown than in general $C \geq (2\pi)^{-\frac{1}{2}}$.

For the case where $X_n \sim \mathcal{B}(p)$ for all n, $\mu = p$ and $\sigma = \sqrt{p(1-p)}$, and

$$\mathbf{E}(|X_1 - \mu|^{3)}) = p|1-p|^3 + (1-p)|0-p|^3 = p(1-p)[p^2 + (1-p)^2].$$

Therefore,

$$\frac{\mathbf{E}(|X_1 - \mu|^{3)})}{\sigma^3} = \frac{p^2 + (1-p)^2}{\sqrt{p(1-p)}} \tag{A.9}$$

so that for all x

$$|Pr\,[S_n^* \leq x] - \mathcal{N}(x)| \leq C \frac{p^2 + (1-p)^2}{\sqrt{p(1-p)}} \frac{1}{\sqrt{n}}. \tag{A.10}$$

We should note that (A.10) will be useful for studying $n \to \infty$, but it is not a good error estimate. With $p = 0.2$, $n = 10$, $C = 0.7975$ the right hand side of (A.10) is about 0.43, but we know the error is much less.

A.6 Complementary Binomials and Normals

Let us suppose that x is not an integer (as will be the case in practice). Then

$$\begin{aligned}\Phi(x;n,p) &= 1 - \Psi(x;n,p) \\ &= 1 - Pr\,[S_n \leq x] \\ &= 1 - Pr\left[S_n^* \leq \frac{x - np}{\sqrt{np(1-p)}}\right];\end{aligned}$$

and so

$$\begin{aligned}&\left|\Phi(x;n,p) - \mathcal{N}(\frac{np - x}{\sqrt{np(1-p)}})\right| \\ &= \left|1 - \Psi(x;n,p) - [1 - \mathcal{N}(\frac{np - x}{\sqrt{np(1-p)}})]\right|\end{aligned}$$

$$= \left|\Psi(x;n,p) - \mathcal{N}(\frac{np-x}{\sqrt{np(1-p)}})\right|$$

$$= \left|Pr\left[S_n \leq x\right] - \mathcal{N}(\frac{np-x}{\sqrt{np(1-p)}})\right|$$

$$= \left|Pr\left[S_n^* \leq \frac{x-np}{\sqrt{np(1-p)}}\right] - \mathcal{N}(\frac{np-x}{\sqrt{np(1-p)}})\right|$$

$$\leq C\frac{p^2 + (1-p)^2}{\sqrt{p(1-p)}} \frac{1}{\sqrt{n}}.$$

That is,

$$\left|\Phi(x;n,p) - \mathcal{N}(\frac{np-x}{\sqrt{np(1-p)}})\right| \leq C\frac{p^2 + (1-p)^2}{\sqrt{p(1-p)}} \frac{1}{\sqrt{n}}. \quad (A.11)$$

It is this estimate that we can use to show that the CRR model price for a call option (say) converges to the corresponding Black and Scholes pricing formula.

A.7 CRR and the Black and Scholes Formula

By the Berry-Esséen Theorem,

$$\left|\Phi(a;N,p) - \mathcal{N}(\frac{Np-a}{\sqrt{Np(1-p)}})\right| \leq C\frac{p^2 + (1-p)^2}{\sqrt{p(1-p)}} \frac{1}{\sqrt{N}} \quad (A.12)$$

and p also depends on N as

$$p = \frac{\exp(r\Delta t) - \exp(-\sigma\sqrt{\Delta t})}{\exp(+\sigma\sqrt{\Delta t}) - \exp(-\sigma\sqrt{\Delta t})}$$

where $\Delta t = \frac{T}{N}$. Using Taylor approximations (and order notation),

$$\pi = \frac{r - \frac{1}{2}\sigma^2}{2\sigma}\sqrt{\Delta t} + \frac{1}{2} + o(\sqrt{\Delta t}) \quad (A.13)$$

$$\pi' = \frac{r + \frac{1}{2}\sigma^2}{2\sigma}\sqrt{\Delta t} + \frac{1}{2} + o(\sqrt{\Delta t}) \quad (A.14)$$

for large N, so $\pi, \pi' \to 0.5$ as $N \to \infty$. Thus, the right hand side of (A.12) tends to zero as $N \to \infty$. In the CRR application for pricing European calls we use

A The Binomial Distribution

$$a \approx \frac{\ln(\frac{K}{S}) - N\ln(d)}{\ln(\frac{u}{d})} = \frac{\ln(\frac{K}{S}) + N\sigma\sqrt{\Delta t}}{2\sigma\sqrt{\Delta t}} = \frac{\ln(\frac{K}{S})}{2\sigma\sqrt{T}}\sqrt{N} + \frac{N}{2}. \quad (A.15)$$

Then

$$\frac{N\pi - a}{\sqrt{N\pi(1-\pi)}} \approx \frac{\left[\frac{N}{2} + \frac{r - \frac{1}{2}\sigma^2}{2\sigma}\sqrt{T}\sqrt{N}\right] - \left[\frac{\ln(\frac{K}{S})}{2\sigma\sqrt{T}}\sqrt{N} + \frac{N}{2}\right]}{\sqrt{N}\sqrt{\frac{1}{2} \cdot \frac{1}{2}}}$$

$$= \frac{r - \frac{1}{2}\sigma^2}{\sigma}\sqrt{T} + \frac{\ln(\frac{S}{K})}{\sigma\sqrt{T}}$$

$$= \frac{\ln(\frac{S}{K}) + (r - \frac{1}{2}\sigma^2)T}{\sigma\sqrt{T}}$$

$$= d_2.$$

Therefore,

$$\phi(a; N, \pi) \to \mathcal{N}(d_2)$$

as $N \to \infty$. In precisely the same way

$$\phi(a; N, \pi') \to \mathcal{N}(d_1)$$

as $N \to \infty$, where

$$d_1 = \frac{\ln(\frac{S}{K}) + (r + \frac{1}{2}\sigma^2)T}{\sigma\sqrt{T}}.$$

In conclusion

$$\begin{aligned} C(0,0) &= S\Phi(a; N, \pi') - KR^{-N}\Phi(a; N, \pi) \\ &\to S\mathcal{N}(d_1) - Ke^{-rT}\mathcal{N}(d_2) \end{aligned} \quad (A.16)$$

as $N \to \infty$. Consequently,

$$C(0) = S\mathcal{N}(d_1) - Ke^{-rT}\mathcal{N}(d_2), \quad (A.17)$$

which is the Black and Scholes formula.

B

An Application of Linear Programming

One of the advantages of considering binomial models is that they provide a complete model for assets. This means that all payoffs can be hedged and priced in the way we have described. One could ask: Why not consider **trinomial** models? In this appendix we show that we can no longer obtain unique prices for contingent claims, but we can obtain a **bid-ask spread**. Prices falling within this spread are consistent with no-arbitrage, but prices falling outside this spread lead to arbitrage opportunities. In this situation we can no longer talk about replication, but we can still obtain what are called **superreplication** and **subreplication**. We shall use **linear programming** to study these situations.

Let us consider a market in which we have a stock and a bank account at times $t = 0$ and $t = 1$. At time $t = 1$ there are now three, (or more!), states of the world, which we label $\uparrow$, $\rightarrow$ and $\downarrow$. We will write $S(0) = S$ and we have the prices $S(1,\uparrow) = uS$, $S(1,\rightarrow) = mS$, $S(1,\downarrow) = dS$ at time $t = 1$. We will also assume that \$1 (or one unit of local currency) will be worth \$$R$ in all states at $t = 1$. That is, it is a riskless asset.

Consider a (European) call option written on this stock with strike price K expiring at $t = 1$. We seek its present value $C(0)$ given its value at $t = 1$, which is (in terms of the stock price):

$$C(1) = [S(1) - K]^+,$$

which is really three equations

$$C(1,\uparrow) = [S(1,\uparrow) - K]^+$$
$$C(1,\rightarrow) = [S(1,\rightarrow) - K]^+$$
$$C(1,\downarrow) = [S(1,\downarrow) - K]^+.$$

B An Application of Linear Programming

If we try our replicating portfolio method we would seek numbers x, y so that

$$x \cdot S(1, \uparrow) + y \cdot R = [S(1, \uparrow) - K]^+$$
$$x \cdot S(1, \rightarrow) + y \cdot R = [S(1, \rightarrow) - K]^+$$
$$x \cdot S(1, \downarrow) + y \cdot R = [S(1, \downarrow) - K]^+.$$

This is the same as

$$x \cdot u \cdot S + y \cdot R = [u \cdot S - K]^+$$
$$x \cdot m \cdot S + y \cdot R = [m \cdot S - K]^+$$
$$x \cdot d \cdot S + y \cdot R = [d \cdot S - K]^+.$$

Once x and y have been obtained (if they exist), then

$$C(0) = x \cdot S(0) + y.$$

However, a system of three equations in two unknowns does not always have a solution. In fact, we can give an example to show this. Suppose that $S(0) = 80$ and we have the prices $S(1, \uparrow) = 85$, $S(1, \rightarrow) = 80$, $S(1, \downarrow) = 75$, $R = 1.01$, and $K = 78$. The three equations are now

$$85 \cdot x + 1.01 \cdot y = 7$$
$$80 \cdot x + 1.01 \cdot y = 2$$
$$75 \cdot x + 1.01 \cdot y = 0.$$

It is easy to see that the unique solution of the first two equations does not solve the third.

B.1 Incomplete Markets

In the above example we see that the risky asset (the stock) and the riskless asset (cash, bank account) do not **span** all possible assets. We say that our market model of the two assets is therefore incomplete. Binomial models are always complete. This explains their popularity. Assets that can be replicated are called **attainable claims**. In incomplete markets, the attainable claims are priced in the usual way and have a unique price. For nonattainable claims (such as the call option in the example above) there is not a unique price but a range of (non arbitrage) prices. This explains in part the existence of **bid-ask spreads** for prices of many assets. Only theoretical models that approximate reality are ever complete, so **bid-ask spreads** are a fact of life. We consider two ways of dealing with incomplete markets.

B.2 Solutions to Incomplete Markets

There are various approaches to incomplete markets. Here are two of them.

(a) Complete the Market

Add some extra tradeable assets into the market model for which you know the present value. The extra tradeable assets must be such that no arbitrage occurs. An example of this could be to add in some options on the underlying stock and to use its market value. Suppose you knew that the price of a $K = 80$ call was $2, then we could use this as a third asset and complete the market. We can now price the $K = 78$ call as follows. Consider a portfolio with x stock, $\$y$ in cash and z of the $K = 80$ calls. We then solve (the replicating equations)

$$85 \cdot x + 1.01 \cdot y + 5 \cdot z = 7$$
$$80 \cdot x + 1.01 \cdot y + 0 \cdot z = 2$$
$$75 \cdot x + 1.01 \cdot y + 0 \cdot z = 0.$$

Solving gives $x = 0.4$, $y = -29.7029703\ldots$, $z = 0.6$. The $K = 78$ call value is then the value at $t = 0$ of this portfolio:

$$80 \cdot x + 1 \cdot y + 2 \cdot z = \$3.4970297\ldots = \$3.50$$

to the nearest cent.

(b) Superreplication

Let us look at matters from the call writer's perspective. At time $t = 1$ he has a liability of $(7, 2, 0)$, depending on which of the three states occurs. So she must consider a portfolio that covers her liabilities, viz.,

$$85 \cdot x + 1.01 \cdot y \geq 7$$
$$80 \cdot x + 1.01 \cdot y \geq 2 \qquad \text{(B.1)}$$
$$75 \cdot x + 1.01 \cdot y \geq 0,$$

for which the writer's cost at $t = 0$ will be $80 \cdot x + y$. The writer will choose x, y to solve (B.1) and to minimize $80 \cdot x + y$, else another person could write the same call at a cheaper price. Therefore, consider the linear programming problem:

$$\min_{x,y} [80 \cdot x + y]$$

such that

$$85 \cdot x + y \cdot 1.01 \geq 7$$
$$80 \cdot x + y \cdot 1.01 \geq 2 \qquad \text{(B.2)}$$
$$75 \cdot x + y \cdot 1.01 \geq 0.$$

Call the minimum C^*, and suppose this minimum is attained at x^*, y^*.

We now find x*, y* and C*.

Let us note that the linear programme (B.2) is feasible. We can take $x = 1$ and $y = 0$. This choice satisfies the constraints (B.1).

The optimal solution can be found graphically (in this case) or by a number of computer packages which have optimization routines. In MS-EXCEL, SOLVER can be used. Thus

$$x^* = 0.7$$
$$y^* = -51.9802$$
$$C^* = 0.7 \times 80 + -51.9802 = \$4.02.$$

C^* and non arbitrage.

If this call were found to be selling in the market at a price $C > C^*$, then there would be an arbitrage opportunity. This is because $C > C^*$ is the same as

$$C - x^* \cdot S - y^* > 0. \qquad \text{(B.3)}$$

At $t = 0$ short sell the call, buy x^* stock, borrow $-y^*$ cash.

At $t = 1$ (the expiry date) cash settle the call, sell the stock, pay back the loan. This gives a net position

$$-(S(1) - K)^+ + x^* \cdot S(1) + 1.01 \cdot y^* \geq 0. \qquad \text{(B.4)}$$

(recall $y^* < 0$). The inequality (B.4) follows because (x^*, y^*) satisfies the three inequalities in (B.2). Therefore, there is an arbitrage opportunity. There is a positive profit at $t = 0$ and no unfunded liabilities at $t = 1$.

Remark B.1. The writer (seller) of a call is interested in the superreplication problem. He needs to satisfy (B.1) in the cheapest way, so that no other writer of calls can undercut this price. We obtain C^*, the asking price—the smallest price at which the writer can sell.

(c) Subreplication

We can also look at the buyer's perspective. Let x_*, y_* and C_* solve the linear programming problem:

$$\max_{x,y} [80 \cdot x + y]$$
$$85 \cdot x + y \cdot 1.01 \leq 7$$
$$80 \cdot x + y \cdot 1.01 \leq 2 \qquad (B.5)$$
$$75 \cdot x + y \cdot 1.01 \leq 0.$$

A buyer will not wish to pay more than C_* for the call because at such a price one of the constraints in (B.5) would be violated, and it would be better to buy the portfolio than to buy the call.

If the price of a call is $C < C_*$, we can also generate an arbitrage starting with

$$80 \cdot x_* + y_* - C > 0$$

and argue as above. The details are left to the reader.

Remark B.2. The **bid-ask** spread has been calculated. It is $[C_*, C^*]$. Bid-ask spreads can also occur in binomial models if we put constraints on the possible values of a hedge ratio (for example) or if we put restrictions on the dates when we can adjust the hedge, (as we described in Section 5. It is easy to adapt the ideas of this appendix to study these situations.

If we do not have a complete market, no arbitrage conditions put restrictions on the possible range of prices (the bid-ask spread). Other conditions or assumptions can be used to narrow this spread. One such condition is the assumption that "good deals" are not possible, or at least such deals disappear very quickly. This conditions puts restrictions on how large the ratio of the expected excess return over the volatility of a return can be. (The excess asset return is the difference between an asset return and the risk-free interest rate.)

B.3 The Duality Theorem of Linear Programming

Here is a statement of the Kuhn-Tucker theorem:

Theorem B.3 (The Kuhn-Tucker Theorem). *Let $x^* \in \mathbf{R}^n$ be a solution of the optimisation problem:*

$$\max_x f(x)$$
$$\text{subject to} \quad g_i(x) \leq 0, \quad i = 1, 2, \ldots, m.$$

Here f, g_i $(i = 1, 2, \ldots)$ are sufficiently smooth functions. Then there exists $\lambda^* \in \mathbf{R}^m$ so that (x^*, λ^*) satisfy:

1. $\frac{\partial f}{\partial x_j}(x^*) + \sum_{i=1}^m \lambda_i^* \frac{\partial g_i}{\partial x_j}(x^*) = 0$
2. $\lambda_i^* \leq 0$, $i = 1, \ldots, m$
3. $g_i(x^*) \leq 0$, $i = 1, \ldots, m$
4. $\lambda_i^* g_i(x^*) = 0$, $i = 1, \ldots, m$.

Assuming Theorem B.3 is true, we now give a proof of the following result.

Corollary B.4. *Suppose f is concave and the g_i are convex, $i = 1, 2, \ldots, m$. Then if (x^*, λ^*) satisfies conditions 1.–4. of Theorem B.3, then x^* solves the maximization problem of $f(x)$ subject to the given conditions.*

Proof. Let x satisfy $g_i(x) \leq 0$, $i = 1, \ldots, m$. Then

$$f(x) \leq f(x^*) + \sum_{j=1}^n (x_j - x_j^*) \frac{\partial f}{\partial x_j}(x^*)$$

because f is concave. Then the right expression is

$$= f(x^*) - \sum_{j=1}^n (x_j - x_j^*) \left(\sum_{i=1}^m \lambda_i^* \frac{\partial g_i}{\partial x_j}(x^*) \right)$$

$$= f(x^*) - \sum_{i=1}^m \lambda_i^* \left(\sum_{j=1}^n (x_j - x_j^*) \frac{\partial g_i}{\partial x_j}(x^*) \right).$$

However, as each g_i is convex,

$$g_i(x) \geq g_i(x^*) + \sum_{j=1}^n (x_j - x_j^*) \frac{\partial g_i}{\partial x_j}(x^*).$$

Then $-\lambda_i^* \geq 0$ gives

$$f(x) \le f(x^*) - \sum_{i=1}^{n} \lambda_i^*[g_i(x) - g_i(x^*)]$$

$$= f(x^*) - \sum_{i=1}^{n} \lambda_i^* g_i(x) \quad \text{by (4)}$$

$$\le f(x^*) \quad \text{by (2) and } g_i(x) \le 0 \quad \forall i.$$

Therefore x^* is optimal. □

Application 1 *Suppose $b \in \mathbf{R}^n$ and A is an $m \times n$ matrix. We characterize the solution of the linear programming problem:*

$$\min_{\mathbf{x}} \mathbf{b}^T \mathbf{x}$$
$$\mathbf{x} \in \mathbf{R}^n, A\mathbf{x} \ge \mathbf{c}, \quad \mathbf{c} \in \mathbf{R}^m.$$

Let us assume that this linear programming (LP) problem has a solution x^*. For this it suffices to show feasibility (there is an $\mathbf{x} \in \mathbf{R}^n$ with $A\mathbf{x} \ge \mathbf{c}$) and boundedness (there is an $m > -\infty$ such that for all $\mathbf{x}$ with $A\mathbf{x} \ge \mathbf{c}, \mathbf{b}^T \mathbf{x} \ge m$). Applying the Kuhn-Tucker theorem to

$$f(x) = -\sum_{j=1}^{n} b_j x_j$$

$$g_i(x) = c_i - \sum_{j=1}^{n} A_{ij} x_j, \quad i = 1, \ldots, m$$

we have that there is a $\lambda^* \in \mathbf{R}^m$ so that:

1. $-b_j + \sum_{i=1}^{m} \lambda_i^*[-A_{ij}] = 0$,
2. $\lambda^* \le 0$, $i = 1, \ldots, m$,
3. $A\mathbf{x}^* \ge \mathbf{c}$,
4. $\lambda_i^*[c_i - \sum_{j=1}^{n} A_{ij} x_j^*] = 0$, $i = 1, \ldots, m$.

Let $y^* = -\lambda^*$. Then $y^* \in \mathbf{R}^m$ and

1.' $A^T \mathbf{y}^* = \mathbf{b}$,
2.' $\mathbf{y}^* \ge \mathbf{0}$,
3.' $A\mathbf{x}^* \ge \mathbf{c}$,
4.' $y_i^*[c_i - \sum_{j=1}^{n} A_{ij} x_j^*] = 0$, $i = 1, \ldots, m$.

4.' is called the *complementary slackness condition*.

Theorem B.5 (Duality Theorem). *Consider the dual linear programme*

$$\max_{\mathbf{y}} \mathbf{c}^T \mathbf{y}$$
$$\text{subject to} \quad A^T \mathbf{y} = \mathbf{b}$$
$$\text{and} \quad \mathbf{y} \geq \mathbf{0}.$$

Then

1. $\mathbf{y}^*$ *is a solution to the dual LP*,
2. $\mathbf{b}^T \mathbf{x}^* = \mathbf{c}^T \mathbf{y}^*$

Proof. 2. follows from 1.-4. of Application 1, for if we sum equations 1.' – 4.' over i, then

$$\mathbf{c}^T \mathbf{y}^* = \mathbf{y}^{*T} A \mathbf{x}^* = (A^T \mathbf{y}^*)^T \mathbf{x}^* = \mathbf{b}^T \mathbf{x}^*.$$

To prove 1., let $A\mathbf{y} = \mathbf{b}, \mathbf{y} \geq \mathbf{0}$. Then

$$\begin{aligned}\mathbf{c}^T(\mathbf{y}^* - \mathbf{y}) &= \mathbf{c}^T \mathbf{y}^* - \mathbf{c}^T \mathbf{y} \\ &\geq \mathbf{c}^T \mathbf{y}^* - [A\mathbf{x}^*]^T \mathbf{y} \\ &= \mathbf{c}^T \mathbf{y}^* - (\mathbf{x}^*)^T A \mathbf{y} \\ &= \mathbf{c}^T \mathbf{y}^* - (\mathbf{x}^*)^T \mathbf{b} \\ &= \mathbf{c}^T \mathbf{y}^* - \mathbf{b}^T \mathbf{x}^* \\ &= 0 \quad \text{by 2.}\end{aligned}$$

Therefore $\mathbf{y}^*$ is optimal. □

Corollary B.6. *Suppose now that there exists* $\mathbf{y}_0 \gg \mathbf{0}$ *(that is, $y_{0i} > 0$ for all $i > 0$) with $A^T \mathbf{y}_0 = \mathbf{b}$. Then*

$$\mathbf{c}^T \mathbf{y}^* = \sup\{\mathbf{c}^T \mathbf{y} |\ A^T \mathbf{y} = \mathbf{b}, \mathbf{y} \gg \mathbf{0}\}.$$

Proof. Clearly $\mathbf{c}^T \mathbf{y}^* \geq \sup\{\mathbf{c}^T \mathbf{y} |\ A^T \mathbf{y} = \mathbf{b}, \mathbf{y} \gg \mathbf{0}\}$.
Consider ϵ such that $0 < \epsilon \leq 1$. Then $\mathbf{y}_\epsilon = (1 - \epsilon)\mathbf{y}^* + \epsilon \mathbf{y}_0 \gg \mathbf{0}$ for all such ϵ. Then

$$\mathbf{y}_\epsilon = (1 - \epsilon)\mathbf{c}^T \mathbf{y}^* + \epsilon \mathbf{c}^T \mathbf{y}_0 \to \mathbf{c}^T \mathbf{y}^* \quad \text{as } \epsilon \to 0^+.$$

Hence, equality holds. □

Example B.7 (Superreplication). Write

$$A = \begin{bmatrix} R & S_1(1,\omega_1) & \cdots\cdots & S_N(1,\omega_1) \\ R & S_1(1,\omega_2) & \cdots\cdots & S_N(1,\omega_2) \\ \cdots & \cdots & \cdots\cdots & \cdots \\ R & S_1(1,\omega_m) & \cdots\cdots & S_N(1,\omega_m) \end{bmatrix}$$

$$\mathbf{c} = \begin{bmatrix} c(1,\omega_1) \\ \cdots \\ c(1,\omega_m) \end{bmatrix} \qquad \mathbf{x} = \begin{bmatrix} H_0 \\ H_1 \\ \cdots \\ H_N \end{bmatrix} \qquad \mathbf{b} = \begin{bmatrix} 1 \\ S_1(0) \\ \cdots \\ S_N(0) \end{bmatrix}.$$

We seek $\mathbf{x}^* \in \mathbf{R}^{N+1}$ so that $A\mathbf{x}^* \geq \mathbf{c}$ and $\mathbf{b}^T \mathbf{x}^*$ is minimal.

$$\mathbf{b}^T \mathbf{x}^* = \sup\{\mathbf{c}^T \mathbf{y}^* \mid A^T \mathbf{y} = \mathbf{b}, \mathbf{y} \gg \mathbf{0}\}.$$

As there is no arbitrage, there is a $\mathbf{y} \gg 0$ with $A^T \mathbf{y}_0 = \mathbf{b}$. In fact $\mathbf{y}_0 = \frac{\pi}{R}$ for some risk-neutral probability π.

Write $\mathcal{P}$ for the set of risk-neutral probabilities. Then

$$\mathbf{b}^T \mathbf{x}^* = \sup_{\pi \in \mathcal{P}} \left\{ \frac{1}{R} \sum_{i=1}^m \pi_i c(1,\omega_i) \right\} = C^*.$$

In the same way

$$C_* = \inf_{\pi \in \mathcal{P}} \left\{ \frac{1}{R} \sum_{i=1}^m \pi_i c(1,\omega_i) \right\}.$$

B.4 The First Fundamental Theorem of Finance

We shall work in the one time step binomial model where $S(0)$ becomes either $S(1,\uparrow)$ or $S(1,\downarrow)$ at time 1 and \$1 becomes \$R.

Remark B.8. This argument can be easily generalized to $N+1$ assets (1 riskless, N risky, say), and the two states at time $t=1$ replaced by K states with little change. We would then be working in at most $N+1$-dimensional subspace of $\mathbf{R}^{K+1}$.

Let

$$\mathcal{M} = \left\{ \begin{pmatrix} -H_0 - H_1 S(0) \\ H_0 R + H_1 S(1,\uparrow) \\ H_0 R + H_1 S(1,\downarrow) \end{pmatrix} ; H_0, H_1 \in \mathbf{R} \right\}$$

B An Application of Linear Programming

$$K = \left\{ \begin{pmatrix} x_1 \\ x_2 \\ x_3 \end{pmatrix} ; x_1, x_2, x_3 \geq 0 \right\}$$

and

$$K_1 = \left\{ \begin{pmatrix} x_1 \\ x_2 \\ x_3 \end{pmatrix} \in K; x_1 + x_2 + x_3 = 1 \right\}.$$

Then no arbitrage is equivalent to $K \cap M = \{\mathbf{0}\}$. This may be represented graphically as in Figure B.1.

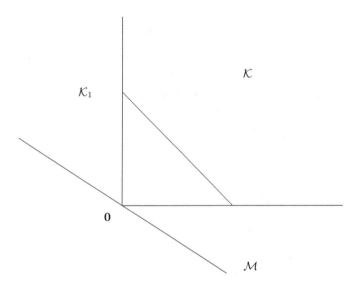

Fig. B.1. Representation of a finite state market.

Lemma B.9. *Assume* $K \cap M = \{\mathbf{0}\}$. *Let* $\mathbf{a} \in K_1$, $\mathbf{m} \in M$ *so that*

$$\|a - m\| = \inf\{\|x - y\| \mid \quad \mathbf{x} \in M, y \in K_1\}$$

and let $\mathbf{p} = \mathbf{a} - \mathbf{m}$. *Clearly,* $\|p\| > 0$. *Then* $\mathbf{p} \perp M$.

B.4 The First Fundamental Theorem of Finance   259

Proof. Let $\mathbf{x} \in \mathcal{M}$ and set

$$f(\epsilon) = \|\mathbf{a} - (\mathbf{m} + \epsilon \mathbf{x})\|^2, \quad \epsilon \in \mathbf{R}.$$

Then $f(\epsilon)$ has a minimum at $\epsilon = 0$, so $f'(0) = 0$. That is

$$f(\epsilon) = \|\mathbf{a} - \mathbf{m}\|^2 - 2\epsilon \langle \mathbf{a} - \mathbf{m}, \mathbf{x} \rangle + \epsilon^2 \|\mathbf{x}\|^2.$$

So $f'(0) = 0$ implies $\langle \mathbf{p}, \mathbf{x} \rangle = \langle \mathbf{a} - \mathbf{m}, \mathbf{x} \rangle = 0$.

□

Lemma B.10. $\mathbf{p} \gg 0$ *(that is, all components of $\mathbf{p}$ are strictly positive)*.

Proof. Let $\mathbf{e} \in \mathcal{K}_1$, $0 \leq \epsilon \leq 1$, and write

$$g(\epsilon) = \|\epsilon \mathbf{e} + (1-\epsilon)\mathbf{a} - \mathbf{m}\|^2.$$

Clearly, $\epsilon \mathbf{e} + (1-\epsilon)\mathbf{a} \in \mathcal{K}_1$ whenever $0 \leq \epsilon \leq 1$. So g has a minimum on $[0,1]$ when $\epsilon = 0$, which implies $g'(0) \geq 0$.
Now

$$g(\epsilon) = \|\mathbf{p}\|^2 - 2\epsilon \langle \mathbf{e} - \mathbf{a}, \mathbf{p} \rangle + \epsilon^2 \|\mathbf{e} - \mathbf{a}\|^2$$

and $g'(0) \geq 0$ implies $\langle \mathbf{e} - \mathbf{a}, \mathbf{p} \rangle \geq 0$ and so

$$\langle \mathbf{e}, \mathbf{p} \rangle \geq \langle \mathbf{a}, \mathbf{p} \rangle \tag{B.6}$$
$$= \langle \mathbf{a} - \mathbf{m} + \mathbf{m}, \mathbf{p} \rangle \tag{B.7}$$
$$= \|\mathbf{p}\|^2 + \langle \mathbf{m}, \mathbf{p} \rangle \tag{B.8}$$
$$= \|\mathbf{p}\|^2 \tag{B.9}$$
$$> 0 \quad \text{by Lemma B.9.} \tag{B.10}$$

Putting $\mathbf{e} = (1,0,0), (0,1,0), (0,0,1)$, we obtain $p_1, p_2, p_3 > 0$. (In fact, $p_i \geq \|\mathbf{p}\|^2$ for $i = 1, 2, 3$).

Lemma B.11. *Let* $\pi_1 = \frac{Rp_2}{p_1}, \pi_2 = \frac{Rp_3}{p_1}$. *Then*

1. $\pi_1 > 0, \pi_2 > 0$,
2. $\pi_1 + \pi_2 = 1$, *and*
3. $RS(0) = \pi_1 S(1, \uparrow) + \pi_2 S(1, \downarrow)$.

Proof. $\mathbf{p} \perp (-1, R, R) \in \mathcal{M}$ (with $H_0 = 1, H_1 = 0$) implies that

$$p_1 = R[p_2 + p_1],$$

from which we claim that 2. and 3. follow. To prove 3.,

$$\mathbf{p} \perp (-S(0), S(1,\uparrow), S(1,\downarrow)) \in \mathcal{M}, \quad (H_0 = 1, H_1 = 0),$$

implies that

$$p_1 S(0) = p_2 S(1,\uparrow) + p_3 S(1,\downarrow).$$

This gives

$$RS(0) = \frac{Rp_2}{p_1} S(1,\uparrow) + \frac{Rp_3}{p_1} S(1,\downarrow)$$
$$= \pi_1 S(1,\uparrow) + \pi_2 S(1,\downarrow).$$

□

We have shown that no arbitrage implies there exists (π_1, π_2), called risk-neutral probabilities, so that

$$\begin{aligned} &\pi_1 > 0, \pi_2 > 0, \\ &\pi_1 + \pi_2 = 1, \quad \text{and} \\ &RS(0) = \pi_1 S(1,\uparrow) + \pi_2 S(1,\downarrow). \end{aligned} \quad (B.11)$$

The converse also holds, for $(R, \pi_1, \pi_2) \perp \mathcal{M}$ and $(R, \pi_1, \pi_2) \cdot \mathbf{y} > 0$ for all $\mathbf{y} \in \mathcal{K} \sim \{\mathbf{0}\}$ implies that $\mathcal{K} \cap \mathcal{M} = \{\mathbf{0}\}$.

This equivalence is the first fundamental theorem of finance.

Corollary B.12. *In a binomial model there is no arbitrage if and only if*

$$S(1,\downarrow) < RS(0) < S(1,\uparrow). \quad (B.12)$$

(We assume here $S(1,\downarrow) < S(1,\uparrow)$.)

Proof. There is no arbitrage if and only if there exists (π_1, π_2) so that Equations (B.11) holds. This is the case if and only if inequality B.12) holds. In fact (B.11) implies

$$S(0)R = \pi_1 S(1,\uparrow) + \pi_2 S(1,\downarrow)$$
$$= S(1,\uparrow) + \pi_2(S(1,\downarrow) - S(1,\uparrow)) < S(1,\uparrow)$$

and

$$S(0)R = \pi_1(S(1,\uparrow) - S(1,\downarrow)) + S(1,\downarrow) > S(1,\downarrow) .$$

Conversely, (B.12) implies

$$\pi_1 = \frac{S(0)R - S(1,\downarrow)}{S(1,\uparrow) - S(1,\downarrow)} > 0$$
$$1 - \pi_1 = \frac{S(1,\uparrow) - S(0)R}{S(1,\uparrow) - S(1,\downarrow)} > 0$$

with π_1, π_2 satisfying (B.11).

B.5 The Duality Theorem

Super- and Subreplication provide the motivation for using the duality theory of linear programming. We shall now illustrate this in a three-state model.

Recall

$$C^* = \inf_{H_0, H_1} \left\{ H_0 + H_1 S(0) \,\middle|\, \begin{array}{l} H_0 R + H_1 S(1,\uparrow) \geq C(1,\uparrow) \\ H_0 R + H_1 S(1,\rightarrow) \geq C(1,\rightarrow) \\ H_0 R + H_1 S(1,\downarrow) \geq C(1,\downarrow) \end{array} \right\} \quad (B.13)$$

and

$$C_* = \inf_{H_0, H_1} \left\{ H_0 + H_1 S(0) \,\middle|\, \begin{array}{l} H_0 R + H_1 S(1,\uparrow) \leq C(1,\uparrow) \\ H_0 R + H_1 S(1,\rightarrow) \leq C(1,\rightarrow) \\ H_0 R + H_1 S(1,\downarrow) \leq C(1,\downarrow) \end{array} \right\} .$$

Remark B.13. The linear program in model (B.13) is feasible for we can choose $H_1 = 0$ and $H_0 R \geq \max\{C(1,\uparrow), C(1,\rightarrow), C(1,\downarrow)\}$. If the market does not have a riskless asset, but one with positive value in each state at $t = 1$, the LP is still feasible.

Remark B.14. The assumption of no arbitrage implies $H_0 + H_1 S(0)$ is bounded below. In fact no arbitrage implies there exists $(\pi_1, \pi_2, \pi_3) \gg \mathbf{0}$ with $\pi_1 + \pi_2 + \pi_3 = 1$ and

$$\pi_1 S(1, \uparrow) + \pi_2 S(1, \rightarrow) + \pi_3 S(1, \downarrow) = RS(0).$$

Therefore, multiplying the equations by $\pi_i, i = 1, 2, 3$ and summing gives

$$H_0 R + H_1 RS(0) \geq \pi_1 C(1, \uparrow) + \pi_2 C(1, \rightarrow) + \pi_3 C(1, \downarrow)$$
$$\geq \min\left[C(1, \uparrow), C(1, \rightarrow), C(1, \downarrow)\right],$$

independently of (π_1, π_2, π_3). Hence, the claim follows.

Remark B.15. Remarks B.13 and B.14 imply that C^* (and C_*) exist.

Write

$$A = \begin{bmatrix} R & S(1,\uparrow) \\ R & S(1,\rightarrow) \\ R & S(1,\downarrow) \end{bmatrix} \qquad \mathbf{c} = \begin{bmatrix} C(1,\uparrow) \\ C(1,\rightarrow) \\ C(1,\downarrow) \end{bmatrix}$$

and

$$\mathbf{x} = \begin{bmatrix} H_0 \\ H_1 \end{bmatrix} \qquad \mathbf{b} = \begin{bmatrix} 1 \\ S(0) \end{bmatrix}.$$

Then

$$C^* = \inf_{H_0, H_1} \left\{ \mathbf{b}^T \mathbf{x} \,|\, A\mathbf{x} \geq \mathbf{c} \right\}. \tag{B.14}$$

Remark B.16. It should be clear that we could discuss this analysis for $N + 1$ assets (1 riskless, N risky assets) and for K states at $t = 1$.
In that case, A will be a $K \times (N + 1)$ matrix, etc. . .

Let $\mathcal{P}$ be the set of all risk-neutral probabilities. Then $\pi \in \mathcal{P}$ if and only if $\pi \gg 0$ and $A^T \pi = R\mathbf{b}$. If there is no arbitrage, then $\mathcal{P} \neq \emptyset$.

Duality

An introduction to linear programming can be obtained from Chvátal's book [16]. The dual linear programming problem for is

$$(P) \begin{cases} \inf_x \mathbf{b}^T \mathbf{x} \\ \quad A\mathbf{x} \geq \mathbf{c} \end{cases}$$

$$(D) \begin{cases} \max_y \mathbf{c}^T \mathbf{y} \\ \quad A^T \mathbf{y} = \mathbf{b} \\ \quad \mathbf{y} \gg \mathbf{0}. \end{cases}$$

Since (P) is feasible, (D) is bounded; since (P) is bounded, (D) is feasible and so (D) has an optimal solution $\mathbf{y}^*$. Further

$$\mathbf{b}^T \mathbf{x}^* = \mathbf{c}^T \mathbf{y}^*,$$

where $\mathbf{x}^*$ is an optimal solution of (P). This follows from the Kuhn-Tucker theorem and is referred to as the **Duality theorem** of linear programming.

By no arbitrage, there exists $\mathbf{y}_0 \gg 0$ with $A^T \mathbf{y}_0 = \mathbf{b}$, for we can choose $y_0 = \frac{\pi}{R}$ with $\pi \in \mathcal{P}$.

Remark B.17. In most of the examples that we have discussed we made one of the assets riskless. This ensures that the linear programme in B.13 are feasible.

Example B.18. Consider now an example of a market determined by two risky assets S_1 and S_2 with data

$$S_1(0) = 0, \ S_1(1,\uparrow) = 1, \ S_1(1,\rightarrow) = -1, \ S_1(1,\downarrow) = 0$$
$$S_2(0) = 1, \ S_2(1,\uparrow) = 1, \ S_12(1,\rightarrow) = -1, \ S_2(1,\downarrow) = 3$$

A claim C is attainable in this market if and only if $C(1,\uparrow) + C(1,\rightarrow) = 0$. So C defined by

$$C(1,\uparrow) = 1, \ C(1,\rightarrow) = 1, \ C(1,\downarrow) = 0$$

is not attainable. Furthermore it is easy to show that there is no H_1 and H_2, so that (three inequalities)

$$H_1 S_1(1) + H_2 S_2(1) \geq C(1) .$$

This means that C **cannot be had at any price**. One can see that the existence of the riskless asset (cash) guarantees the old adage **"everything has a price"**.

B.6 The Second Fundamental Theorem of Finance

This theorem says that in a market with no arbitrage, the market is complete if and only if $\mathcal{P}$ contains just one element.

In the example above (which can be easily generalized to $N+1$ assets and K states, say), let us write that if π^1 and $\pi^2 \in \mathcal{P}$, then there is a claim C so that $C^* \neq C_*$ (so the market is incomplete). In fact, if $\pi_j^1 < \pi_j^2$, $C(1,j) = 1, C(1,i) = 0, i = j$, then

$$C_* = \inf_{\pi \in \mathcal{P}} \left\{ \frac{1}{R} \sum_i \pi_i C(1,i) \right\}$$
$$\leq \frac{1}{R}\pi_j^1$$
$$< \frac{1}{R}\pi_j^2$$
$$\leq \sup_{\pi \in \mathcal{P}} \left\{ \frac{1}{R} \sum_i \pi_i C(1,i) \right\}$$
$$= C^*.$$

Suppose that $\mathcal{P}$ is a singleton, then $C^* = C_*$ for every claim C. Suppose C is not attainable. Without loss of generality we may assume $\mathbf{H}^*$ is the optimal solution for C^*, i.e.,

$$C^* = \mathbf{b}^T \mathbf{H}^*,$$
$$A\mathbf{H}^* \geq \mathbf{c} \quad \text{and} \quad (A\mathbf{H}^*)_j > C(1,j) \text{ for some } j.$$

Then if $\pi \in \mathcal{P}$,

$$\pi^T A \mathbf{H}^* > \pi^T \mathbf{C}.$$

Therefore

$$(A^T \pi)^T \mathbf{H}^* > \mathbf{C}^T \pi$$

so

$$R \mathbf{b}^T \mathbf{H}^* > \mathbf{C}^T \pi$$

and

B.6 The Second Fundamental Theorem of Finance

$$\mathbf{b}^T \mathbf{H}^* > \frac{1}{R} \mathbf{C}^T \pi \geq C_*$$

or

$$C^* > C_*,$$

a contradiction.

Attainable Claims

A claim C is attainable if and only if $C_* = C^*$ which occurs if and only if $\mathbf{c}^T \pi$ is independent of π for $\pi \in \mathcal{P}$ (see Pliska, [62]).

The following example shows how one can identify attainable claims.

Suppose

$$S(1,\uparrow) = 120 \qquad C(1,\uparrow) = C_1$$

$$S(1,\rightarrow) = 110 \qquad C(1,\rightarrow) = C_2$$

$$S(1,\downarrow) = 90 \qquad C(1,\downarrow) = C_3$$

$$S(0) = 100 \qquad R = 1.05$$

(π_1, π_2, π_3) is a risk-neutral probability if $\pi_i > 0$, $\pi_1 + \pi_2 + \pi_3 = 1$ and $12\pi_1 + 11\pi_2 + 9\pi_3 = 10.5$. These can be parameterized as

$$\mathcal{P} = \left\{ (\pi_1, \pi_2, \pi_3) \mid \begin{pmatrix} \pi_1 \\ \pi_2 \\ \pi_3 \end{pmatrix} = \begin{pmatrix} 2 \\ -3 \\ 1 \end{pmatrix} t + \begin{pmatrix} -\frac{1}{2} \\ \frac{3}{2} \\ 0 \end{pmatrix} \frac{1}{4} < t < \frac{1}{2} \right\}.$$

C is attainable if and only if $C_1\pi_1 + C_2\pi_2 + C_3\pi_3$ is independent of t. That is,

$$(2C_1 - 3C_2 + C_3)t + (-\frac{1}{2}C_1 + \frac{3}{2}C_2)$$

is independent of t, so we require $2C_1 - 3C_2 + C_3 = 0$.
If $2C_1 - 3C_2 + C_3 > 0$, then

$$C^* = \frac{1}{R}\left[\frac{1}{2}(2C_1 - 3C_2 + C_3) + (-\frac{1}{2}C_1 + \frac{3}{2}C_2)\right]$$

$$= \frac{1}{1.05}\left[\frac{1}{2}C_1 + 2C_3\right]$$

$$C_* = \frac{1}{R}\left[\frac{1}{4}(2C_1 - 3C_2 + C_3) + (-\frac{1}{2}C_1 + \frac{3}{2}C_2)\right]$$

$$= \frac{1}{1.05}\left[\frac{3}{4}C_2 + \frac{1}{2}C_3\right].$$

This example is typical of how C^* and C_* can be computed.

Remark B.19. The existence of the riskless asset is tacitly assumed. Recall Example B.18. $\mathcal{P}$ consists of all $(p_1, p_2, p_3) \gg 0$ so that

$$RS_1(0) = p_1 S_1(1,\uparrow) + p_2 S_1(1,\rightarrow) + p_3 S_1(1,\downarrow)$$
$$RS_2(0) = p_1 S_2(1,\uparrow) + p_2 S_2(1,\rightarrow) + p_3 S_2(1,\downarrow)$$
$$1 = p_1 + p_2 + p_3.$$

This system has a unique solution: $p_1 = p_2 = p_3 = \frac{1}{3}$ and $R = 1$, but the market is incomplete as we have seen. The reader is also referred to [59, chapter 5], for further discussion.

B.7 Transaction Costs

Example B.20. Boyle and Tan [9] discuss a three-time example of option hedging with proportional transaction costs. In our notation we have $t = 0, 1, 2$. If we wish to calculate the ask price for a claim that pays $C(2, j)$ at time $t = 2$ in state j, we should solve the following programming problem: Find the minimum value of

$$H_0(0,0) + H_1(0,0)S(0,0)$$

subject to constraints

$$H_0(0,0)R + H_1(0,0)S(1,j) \geq H_0(1,j) + H_1(1,j)S(1,j) + T(1,j)$$

$$H_0(1,j)R + H_1(1,j)S(2,j+1) \geq C(2, j+1)$$

$$H_0(1,j)R + H_1(1,j)S(2,j) \geq C(2, j)$$

for $j = 0, 1$, where

$$T(1,j) = k|H_1(1,j) - H_1(0,0)|S(1,j)$$

are proportional transaction costs, and $k \geq 0$ indicates the level of transaction costs.

There are 6 constraint equations here. The difficulty is how to deal with the $|H_1(1,j) - H_1(0,0)|$ term. This nonlinear programming problem can be converted into a mixed integer linear programming problem (but with 8 rather than 6 variables). This is achieved as follows.

Write $H_1(1,j) - H_1(0,0) = y_j - z_j$ where $y_j \geq 0$, $z_j \geq 0$ and $y_j \cdot z_j = 0$. Then $|H_1(1,j) - H_1(0,0)| = y_j + z_j$ for $j = 0, 1$. We now obtain a set of linear constraints by using these new variables, but introduce two new nonlinear constraints, $y_0 \cdot z_0 = 0$ and $y_1 \cdot z_1 = 0$.

These last two constraints can be converted into EITHER/OR constraints (see [78, pages 477–480]), for $y_j \cdot z_j = 0$ is the same as EITHER $y_j = 0$ OR $z_j = 0$. These can be treated through introducing integer constraints. These nonlinear programming problems are then solvable with computer packages that handle mixed integer linear programming problems. CPLEX is a well known example of a computer package that meets these needs. Clearly these ideas can be extended to more complicated examples, and CPLEX can handle such problems with thousands of variables.

C

Volatility Estimation

One of the most important variables in option pricing is the volatility of the 'underlying' (as we call the underlying asset). In fact it is important because it is this factor that usually most effects an option price. Practitioners sometimes quote the volatility of an underlying rather than quoting the price of an option. This gives rise to the term "trading volatility". The book by Natenberg [57] is very popular with practitioners and includes a discussion of volatility.

In order to determine the volatility of an underlying, one should have a model for the underlying. Perhaps this is one of the most important tasks in finance—producing and calibrating models for price processes (asset prices, exchange rates, interest rates, and so on). There are many such models now available. One likes to be parsimonious, selecting the simplest model which will reproduce "stylized facts" observed in "the market." There are continuous time models, discrete time models—like the binomial tree models, time series models—like ARCH, GARCH, EGARCH and so on. In this book we have concentrated on binomial tree models. These can be calibrated, and relatively easy algorithms are available for pricing quite complicated derivative assets.

In this section we shall use the model for an underlying (asset) developed by Black and Scholes to obtain their famous European call and put option pricing formulae. We shall calibrate the volatility for this model. This is also the calibration we needed for using the CRR model.

Assume that the underlying asset price S satisfies

$$\ln\left(\frac{S(t_{j+1})}{S(t_j)}\right) = \mu(t_{j+1} - t_j) + \sigma\sqrt{t_{j+1} - t_j} \cdot \epsilon_{j+1} \quad \text{(C.1)}$$

where $0 = t_0 < t_1 < t_2, \ldots$, and the random variables ϵ_j are all independent and identically distributed $\mathcal{N}(0, 1)$.

There are basically two approaches to estimating volatility:

1. historical volatility estimation;

2. implicit or implied volatility estimation.

C.1 Historical Volatility Estimation

In order to estimate the expected return μ and the volatility σ, we assume that we have equally spaced measurements of S. That is, $(t_{j+1} - t_j) = \Delta t$ for each j. These could be daily stock prices, which we will denote by

$$S_0, S_1, S_2, S_3, \ldots, S_N. \tag{C.2}$$

We then take natural logarithms of ratios of consecutive terms. Put $Y_i = \ln\left(\frac{S_i}{S_{i-1}}\right)$ to obtain

$$Y_1, Y_2, Y_3, \ldots, Y_N. \tag{C.3}$$

We can then estimate $\mu \Delta t$ by

$$\hat{\mu} = \frac{1}{N} \sum_{i=1}^{N} Y_i \tag{C.4}$$

and $\sigma^2 \Delta t$ by

$$\hat{\sigma}^2 = \frac{1}{N-1} \sum_{i=1}^{N} (Y_i - \hat{\mu})^2 \tag{C.5}$$

$$= \frac{1}{N-1} \sum_{i=1}^{N} Y_i^2 - \frac{N}{N-1} \hat{\mu}^2. \tag{C.6}$$

We note that $N-1$ is used rather than N. This gives the well known unbiased estimator for $\sigma^2 \Delta t$. Equation (C.6) follows from (C.4), and may simplify the calculation of $\hat{\sigma}^2$.

We now provide an example. This is taken from Hull [37, page 234].

Example C.1. From Table C.1, we compute

$$\sum_{i=1}^{20} Y_i = 0.09531, \quad \sum_{i=1}^{20} Y_i^2 = 0.00333$$

and hence the estimations (using (C.4) and (C.6)):

C.1 Historical Volatility Estimation

Table C.1. Daily stock returns

Day	Stock Price	$\frac{S_i}{S_{i-1}}$	Daily return Y_i
0	20.000		
1	20.125	1.006250	0.006231
2	19.875	0.987578	-0.012500
3	20.000	1.006289	0.006270
4	20.500	1.025000	0.024693
5	20.250	0.987805	-0.012270
6	20.875	1.030864	0.030397
7	20.875	1.000000	0.000000
8	20.875	1.000000	0.000000
9	20.750	0.994012	-0.006010
10	20.750	1.000000	0.000000
11	21.000	1.012048	0.011976
12	21.250	1.011905	0.011834
13	20.875	0.982353	-0.017800
14	20.875	1,000000	0.000000
15	21.250	1.017964	0.017805
16	21.375	1.005882	0.005865
17	21.375	1.000000	0.000000
18	21.250	0.994152	-0.005870
19	21.750	1.023529	0.023257
20	22.000	1.011494	0.011429

$$\hat{\mu} = 0.00476, \quad \hat{\sigma}^2 = 0.000151357$$

or $\hat{\sigma} = 0.00123$. We are not so concerned about $\hat{\mu}$ as this does not enter option pricing. But $\hat{\sigma}$ is an estimator for $\sigma^2 \Delta t$, or the daily volatility. We must decide how time is measured. If time is measured in trading days, and we assume (as does Hull) that there are 252 trading days per year, then $\Delta t = 1/252$ and that the data give an estimate for the volatility per annum (annualized volatility) of $0.00123 \times \sqrt{252} = 0.195$. The estimated volatility is 19.5% per annum. (The standard error of this estimate is $\frac{0.195}{\sqrt{2 \times 20}}$ or 3.1% per annum.)

Remark C.2. 1. What is the daily price? Hull in the example considers the closing stock price on each trading day. Are there other choices? After all, much activity occurs each trading day. Probably our decision of what to do with the estimates will dictate what choices we make.

2. Are there problems counting Friday to Monday as one day?

3. Is it reasonable to assume the volatility is the same over the whole week. If we were to take just Wednesday to Tuesday ratios for the Y_i, would we get a different answer?

4. If we have used trading days to measure time here, then when we use the historical volatility to price an option by the Black and Scholes formula we

should also measure time in trading days. (This should make us conscious of how data are collected and how estimations are made.)

5. As to the number of trading days, this may vary from country to country. Some take 260. Perhaps it should be approximately $365 - 2 \times 52 = 261$.

6. We have assumed here and below that the stock is a non dividend-paying stock. Modifications can be made for dividends —see Hull [ibid., page 235]. Of course, this approach can be used to estimate the volatility of other processes, which are assumed to evolve like those in Table C.1, for example exchange rates, interest rates, and so on. However, for other examples different models to C.1 may be required to capture market features (stylized facts).

C.2 Implied Volatility Estimation

Implied volatility is calculated in a different way. We still assume the model in C.1, and we assume that the Black and Scholes formula for a European call (put) option is correct. For the European call option with time to expiry T, and strike (or exercise) price K, this is

$$C = S\mathcal{N}(d_1) - Ke^{-rT}\mathcal{N}(d_2) \tag{C.7}$$

where

$$d_1 = \frac{\ln\left(\frac{S}{K}\right) + \left(r + \frac{1}{2}\sigma^2\right)T}{\sigma\sqrt{T}}, \quad d_2 = d_1 - \sigma\sqrt{T}, \tag{C.8}$$

where S is the current stock price, r is the risk-free interest rate appropriate to the tenor (duration) of the option, σ the volatility. Of course $\mathcal{N}(x)$ is the usual standard normal distribution function whose values are widely tabulated. Of the inputs into (C.7) S, K, r, T are given, but not σ. The implied volatility is that value of σ so that for a given choice of S, K, r, T, the formula (C.7) produces the call value quoted in the market. Let us now see some ways for calculating this implied volatility. Let us assume that S, K, r, T are given and that

$$f(\sigma) \equiv S\mathcal{N}(d_1) - Ke^{-rT}\mathcal{N}(d_2) - C_m. \tag{C.9}$$

Here C_m denotes the market price of the call option for the choice S, K, r, T. We now seek $\sigma = \sigma_{imp}$ so that

$$f(\sigma_{imp}) = 0. \tag{C.10}$$

C.2 Implied Volatility Estimation

We now describe two standard (popular) methods for solving equation (C.10).

Newton-Raphson Method

This is an iterative method, which usually converges very quickly. We describe how to go from an estimate σ_n to σ_{n+1}. Of course we assume $f(\sigma_n) \neq 0$, else there is not much point proceeding to next iteration, as we already have computed σ_{imp}. For σ_{n+1}, we take

$$\sigma_{n+1} = \sigma_n - \frac{f(\sigma_n)}{f'(\sigma_n)}. \quad (C.11)$$

There are several ways to see why (C.11) is a good approximation. The usual explanation is the following. Look at the graph of the function $y = f(\sigma)$. We can see that this function f is an increasing function of σ. In fact

$$f'(\sigma) = V(\sigma) = S\sigma\sqrt{T}\mathcal{N}'(d_1) > 0 \quad (C.12)$$

where

$$\mathcal{N}'(x) = \frac{1}{\sqrt{2\pi}}\exp(-\frac{1}{2}x^2), \quad (C.13)$$

which is easily computed. Here, let V stand for **vega** which is the derivative of the call price with respect to the volatility (the Greek letter kappa (κ) is also used). Continuing, draw a tangent to the graph of $y = f(\sigma)$ at the point $(\sigma_n, f(\sigma_n))$. This has equation

$$y = f(\sigma_n) + V(\sigma_n)(\sigma - \sigma_n). \quad (C.14)$$

Now compute where this tangent line intersects the σ-axis. Thus

$$0 = f(\sigma_n) + V(\sigma_n)(\sigma - \sigma_n), \quad (C.15)$$

and you will obtain

$$\sigma = \sigma_n - \frac{f(\sigma_n)}{V(\sigma_n)} \quad (C.16)$$

as the value of σ_{n+1}. This is just another way of writing (C.11). By drawing a graph, you can see why σ_{n+1} is a better estimate for σ_{imp} than is σ_n. From a computational point of view, it would be good to program the functions $C = C(\sigma) \equiv S\mathcal{N}(d_1) - Ke^{-rT}\mathcal{N}(d_2)$ and $V(\sigma) \equiv S\sigma\sqrt{T}\mathcal{N}'(d_1)$ with input parameters S, K, r, T. Here is a sample computation which is taken from Chriss [14, page 339].

C Volatility Estimation

Example C.3. Input data is $S = \$100$, $K = \$100$, $T = 2$ years, $r = 3\%$, the market price of the European call option is $C_m = \$8.61$. We begin with an initial guess of $\sigma_1 = 20\%$. We know this is a reasonable value from past experience. We then compute $C(\sigma_1) = \$14.0721$ using the Black and Scholes formula, and so $f(\sigma_1) = C(\sigma_1) - C_m = 14.0721 - 8.61 = 5.4621$. Next we compute the Black and Scholes Vega, $V(\sigma_1) = 53.0007$, and now σ_2,

$$\sigma_2 = \sigma_1 - \frac{f(\sigma_1)}{V(\sigma_1)} = 0.20 - \frac{5.4621}{53.0007} = 0.0969,$$

so the next estimate of volatility is 9.69%.

We proceed to the next iteration. With $\sigma_2 = 9.69$, $C(\sigma_2) = 8.7144$, $V(\sigma_2) = 49.6346$, $f(\sigma_2) = V(\sigma_2) - C_m = 0.1044$, and so

$$\sigma_3 = \sigma_2 - \frac{f(\sigma_2)}{V(\sigma_2)} = 0.0969 - \frac{0.1044}{49.6346} = 0.0948$$

which is 9.48%.

Since $C(\sigma_3) = C(0.0948) = 8.6103$ we could decide to stop now and conclude that a satisfactory estimate for $\sigma_{imp} = 9.48\%$.

The Method of Bisections

This method is also iterative and may produce a value of the implied volatility in more steps than the Newton-Raphson method, but it has the advantage that we do not need an explicit formula for $C(\sigma)$ or its derivative as in the above method.

We start out with an initial guess for the implied volatility, which we will call σ_0, and the compute $f(\sigma_0) = C(\sigma_0) - C_m$. If $f(\sigma_0) > 0$, then we can conclude from the above discussion that $\sigma_0 > \sigma_{imp}$. We now need to find another guess that makes f lower. Set σ_1 to σ_0 reduced by 50%, that is,

$$\sigma_1 = \sigma_0 - \frac{1}{2}\sigma_0 = 0.5\sigma_0$$

We then compute $f(\sigma_1)$ as before. If $f(\sigma_1) > 0$, then we again conclude that $\sigma_1 > \sigma_{imp}$, and we again reduce σ_1 by 50% of the earlier change:

$$\sigma_2 = \sigma_1 - \frac{1}{4}\sigma_0.$$

On the other hand if $f(\sigma_1) < 0$ then we conclude that $\sigma_1 < \sigma_{imp}$, and we increase σ_1 by 50% of the earlier change:

$$\sigma_2 = \sigma_1 + \frac{1}{4}\sigma_0.$$

C.2 Implied Volatility Estimation

We then iterate these steps. We compute σ_k from σ_{k-1} as follows:

$$\sigma_k = \begin{cases} \sigma_{k-1} - \frac{\sigma_0}{2^k} & \text{if } f(\sigma_{k-1}) > 0 \\ \sigma_{k-1} + \frac{\sigma_0}{2^k} & \text{if } f(\sigma_{k-1}) < 0. \end{cases} \tag{C.17}$$

It should be clear that with each step the estimates of the σ_{imp} are getting closer and closer. If we require an accuracy of ϵ for σ_{imp}, then we continue the iterations until

$$\frac{\sigma_0}{2^k} < \epsilon$$

or

$$k > \frac{\ln\left(\frac{\sigma_0}{\epsilon}\right)}{\ln(2)}. \tag{C.18}$$

If we took $\sigma_0 = 20\%$, and $\epsilon = 0.01\%$, then (C.18) means $k > 10.97$, that is, we need at least 11 steps. We now present a sample calculation (see Chriss [ibid., page 333]).

Example C.4. The data are the same as in the previous example using the Newton-Raphson method. We start with $\sigma_0 = 0.20$ as before and choose $\epsilon = 0.0001$. Since the option's actual volatility is not greater than 20% (check that $f(0.20) > 0$), we define σ_1 by

$$\sigma_1 = \sigma_0 - \frac{1}{2}\sigma_0 = 0.10.$$

Now $f(\sigma_1) = 8.87 - 8.61 > 0$, so σ_1 is still too large. Now select σ_2 by

$$\sigma_2 = \sigma_1 - \frac{1}{4}\sigma_0 = 0.10 - \frac{1}{2} \times 0.20 = 0.05,$$

and now $f(\sigma_2) = 6.58 - 8.61 < 0$, which means that σ_2 is too small. Now select σ_3 by

$$\sigma_3 = \sigma_2 + \frac{1}{8}\sigma_0 = 0.05 + \frac{1}{8} \times 0.20 = 0.075,$$

and now $f(\sigma_3) = 7.66 - 8.61 < 0$, which means that σ_3 is too small. Now select σ_4 by

$$\sigma_4 = \sigma_3 + \frac{1}{16}\sigma_0 = 0.075 + \frac{1}{16} \times 0.20 = 0.0875,$$

and now $f(\sigma_4) = 8.25 - 8.61 < 0$, which means that σ_4 is too small. Now select σ_5 by

$$\sigma_5 = \sigma_4 + \frac{1}{32}\sigma_0 = 0.0875 + \frac{1}{32} \times 0.20 = 0.0938, \qquad (*)$$

and now $f(\sigma_5) = 8.56 - 8.61 < 0$, which means that σ_5 is too small. Now select σ_6 by

$$\sigma_6 = \sigma_5 + \frac{1}{64}\sigma_0 = 0.0938 + \frac{1}{64} \times 0.20 = 0.0969,$$

and now $f(\sigma_6) = 8.71 - 8.61 > 0$, which means that σ_6 is too large. Now select σ_7 by

$$\sigma_7 = \sigma_6 - \frac{1}{128}\sigma_0 = 0.0969 - \frac{1}{128} \times 0.20 = 0.0953,$$

and now $f(\sigma_7) = 8.64 - 8.61 > 0$, which means that σ_7 is too large. Now select σ_8 by

$$\sigma_8 = \sigma_7 - \frac{1}{256}\sigma_0 = 0.0953 - \frac{1}{256} \times 0.20 = 0.0945,$$

and now $f(\sigma_8) = 8.60 - 8.61 < 0$, which means that σ_8 is too small. Now select σ_9 by

$$\sigma_9 = \sigma_8 + \frac{1}{512}\sigma_0 = 0.0945 + \frac{1}{512} \times 0.20 = 0.0949,$$

and now $f(\sigma_9) = 8.62 - 8.61 > 0$, which means that σ_8 is too large. Now select σ_{10} by

$$\sigma_{10} = \sigma_9 - \frac{1}{1024}\sigma_0 = 0.0949 - \frac{1}{1024} \times 0.20 = 0.0947,$$

and now $\sigma_{11} - \sigma_{10}$ will be less than 0.0001 and so $\sigma_{imp} = 9.47\%$ to the required accuracy. Let us note that in $(*)$ we did not carry accuracy greater than that required in the final answer. This approach is taken in all further steps.

We summarize the calculations in Table C.2.

Table C.2. Bisection method.

Step i	σ_i	$C(\sigma_i)$
1	10.00%	$8.87
2	5.00%	$6.58
3	7.50%	$7.66
4	8.75%	$8.25
5	9.38%	$8.56
6	9.69%	$8.71
7	9.53%	$8.64
8	9.45%	$8.60
9	9.49%	$8.62
10	9.47%	$8.61

We now make some remarks that apply to both methods.

Remark C.5. 1. Of course both approaches could be applied more generally. But for the Newton-Raphson method, we need an option pricing formula or an algorithm for computing $C = C(\sigma)$ and its derivative $V(\sigma)$. In the method of bisections we do not need the derivative at all. This approach will be convenient if we are trying to "back-out" implied volatilities from American option prices, for example.

2. What we have calculated above is the Black and Scholes implied volatility, but we could also consider computing the CRR implied volatility, when the market prices could be American option prices. Then we must use the method of bisections to calculate the implied volatilities.

3. We have demonstrated the method using market European call option prices. We can clearly do the same from European put prices or combinations of these. We could also use market prices of spreads (spread options are priced as the difference of two option prices) in the same way to compute an implied volatility. This is sometimes done by practitioners.

4. If model (C.1) is valid then the implied volatility calculated in either way just described should give the same answer for any choice of S, K, r, T provided we use the corresponding market price. It is well known that this is not the case. This would seem to throw some doubt on the validity of model (C.1).

5. If we keep S, r, T fixed and compute σ_{imp} for various values of strike K, we can get different answers. This gives rise to the phenomenon called "volatility smiles". We may see this in Table (C.3)

Table C.3. Call prices and implied volatility.

K	Call price-ask	Implied volatility-ask
17.00	3.27	71.70
17.50	2.75	62.50
18.00	2.27	55.20
18.50	1.79	47.90
19.00	1.35	41.70
19.50	0.95	36.20
20.00	0.63	32.40
20.50	0.38	29.20
21.00	0.22	27.60
21.50	0.12	26.50
22.00	0.07	26.60

6. You may say that this example does not exhibit a smile. Usually the lowest value of implied volatilities occurs for the ATM (at-the-money) options.

7. Here is a piece of anecdotal information. Implied volatility estimation should be the same if we use market put prices or market call prices. However, practitioners use out-of-the-money option prices to estimate implied volatilities. So in the example they may use puts to estimate implied volatilities when $K < \$20.14$ and calls when $K > \$20.14$. Why do you think this is done?

8. What is to be done? The famous book by Cox and Rubinstein [19] has a chapter called "How to Use the Black and Scholes Formula". They describe various ways of taking weighted averages of various answers for implied volatility to obtain a value of the volatility that is appropriate for the stock in hand. This number is then used to price options or to generate quotes on options that are not currently being traded, or to price various exotic options on the same stock, which are not traded in the exchanges.

9. Another way around this problem is to plot σ_{imp} versus K and determine a smooth curve that goes through these points. For example, some practitioners might use a spline function to fit these data points. Then, for pricing other options (nonmarket choices of K), use this graph to determine a suitable choice of volatility.

10. Another issue is the following. Which is the best to use—historical volatility estimates or implied volatility estimates? There is quite some debate about this. It is also possible that these estimates could be quite different. After all, the historical volatility uses historical data, whereas the implied volatility uses market information about future volatilities. So perhaps the implied volatilities are more appropriate. However, consider how market prices of options are determined. These are no doubt based on past history as well as on present assessment of the future based on recent history. (We are excluding inside information about the future here.) Perhaps implied volatilities are just another type of "historical" volatility estimation. This raises the interesting issue of price formation in markets (price discovery).

11. Suppose now that we keep S, K, r fixed, and calculate σ_{imp} for various values of T. If model (C.1) is correct then again we should always get the same answer. However, we often obtain different answers. This means that the (annualized) volatilities for different times to expiry are not the same. For example, option prices imply greater volatility close to the expiry of an option. We can then talk about the term structure of volatility, just as we talk about the term structure of interest rates. This is another active area of research.

Further references for calculating implied volatility include Beckers [4], Mayhew [49] and Latane and Rendleman [45].

C.3 Exercises

Exercise C.6. Implement Example C.3.

Exercise C.7. Implement Example C.4.

D

Existence of a Solution

In this note we discuss the existence of a solution to the system

$$1 = \sum_{j=0}^{N} Q_j \tag{D.1}$$

$$\rho \cdot S = \sum_{j=0}^{N} Q_j S(N, j) \tag{D.2}$$

$$\rho \cdot \tilde{C}_i = \sum_{j=0}^{N} Q_j \left[S(N, j) - K_i\right]^+ \quad \text{for } i = 1, 2, \ldots, m \tag{D.3}$$

with $Q_j \geq 0$ for all $j = 0, 1, 2, \ldots, N$ and $\rho = \exp(rT)$. We shall show that (D.1), (D.2), (D.3) has a nonnegative solution for $Q_0, Q_1, \ldots, Q_N$ if and only if there are no arbitrage opportunities in the market of the stock and the various options whose prices appear in these formulae. In order to show this we need Farkas' lemma (1902). (Farkas' name is pronounced in the Hungarian manner as "Farkash").

D.1 Farkas' Lemma

This famous result has many applications in optimization theory and elsewhere.

Lemma D.1 (Farkas). *Let A be an $m \times n$ matrix and $b \in \mathbf{R}^m$. Then the system*

D Existence of a Solution

$$Ax = b \quad x \in \mathbf{R}^n \quad x \geq 0^1 \tag{D.4}$$

has a solution x if and only if the system

$$w'A \leq 0 \quad w \in \mathbf{R}^m \quad w'b > 0 \tag{D.5}$$

has no solution w.[2]

Proof. (a) Let us assume (D.4) has a solution $x \geq 0$. Assume also, if possible, that (D.5) has a solution w. Then

$$0 \geq w'Ax = w'b > 0$$

which is a contradiction. And so (D.5) cannot have a solution.

(b) Let us assume that (D.4) does not have a solution, and let us show that (D.5) then has a solution.

Let

$$S = \{y \in \mathbf{R}^m | y = Ax \text{ with } x \in \mathbf{R}^n \text{ and } x \geq 0\}.$$

Then S is a closed convex subset of $\mathbf{R}^m$ and $b \notin S$.

Let y^* be the point in S which is closest to b. Then [3]

$$\langle b - y^*, y^* - y \rangle \geq 0 \tag{D.6}$$

for all $y \in S$. To see why this is true, set

$$\begin{aligned}
\psi(t) &= \|b - (ty + (1-t)y^*)\|^2 \\
&= \|(b - y^*) + t(y^* - y)\|^2 \\
&= \|b - y^*\|^2 + 2t\langle b - y^*, y^* - y \rangle + t^2\|y^* - y\|^2
\end{aligned}$$

for $0 \leq t \leq 1$. Then from the definition of y^*, ψ has a minimum at $t = 0$, and so $\psi'(0) \geq 0$, from which (D.6) follows.

Let $w = b - y^* \neq 0$. We claim that this choice satisfies (D.5).
In fact, using $0 \in S$ in (D.6),

$$\begin{aligned}
w'b &= \langle b - y^*, b \rangle \\
&= \langle b - y^*, b - y^* \rangle + \langle b - y^*, y^* \rangle
\end{aligned}$$

---

[1] $x = (x_1, x_2, \ldots, x_n)' \geq 0$ means $x_i \geq 0$ for all $i = 1, 2, \ldots, n$.
[2] w' denotes the transpose of the vector w.
[3] If $x, y \in \mathbf{R}^n$, say, then $\langle x, y \rangle = x_1 y_1 + x_2 y_2 + \ldots + x_n y_n$, the scalar (dot) product of x and y.

$$\geq \langle b-y^*, b-y^* \rangle = \|b-y^*\|^2 > 0$$

Also rearranging (D.6) gives

$$\langle b-y^*, y \rangle \leq \langle b-y^*, y^* \rangle \tag{D.7}$$

for all $y \in S$. This is the same as

$$w'Ax \leq \langle b-y^*, y^* \rangle \tag{D.8}$$

for all $x \in \mathbf{R}^n$ with $x \geq 0$. Now choose $x = ne_i$ (integer n times the unit vector e_i[4]). Then

$$(w'A)_i = \frac{1}{n}w'A(ne_i) \leq \frac{1}{n}\langle b-y^*, y^* \rangle, \tag{D.9}$$

and letting $n \to +\infty$, (D.9) implies that

$$(w'A)_i \leq 0.$$

As this can be applied with any $i = 1, 2, \ldots, n$, each component of $w'A$ is less than or equal to zero, which is the same as saying $w'A \leq 0$. We have then shown that w defined above satisfies (D.5). We have thus established that if (D.5) has no solution, then (D.4) must have a solution (for if not ...). □

D.2 An Application to the Problem

We apply Farkas' lemma with

$$A = \begin{pmatrix} 1 & 1 & \cdots & 1 \\ (S(N,0)-K_1)^+ & (S(N,2)-K_1)^+ & \cdots & (S(N,N)-K_1)^+ \\ (S(N,0)-K_2)^+ & (S(N,2)-K_2)^+ & \cdots & (S(N,N)-K_2)^+ \\ \vdots & \vdots & & \vdots \\ (S(N,0)-K_m)^+ & (S(N,2)-K_m)^+ & \cdots & (S(N,N)-K_m)^+ \\ S(N,0) & S(N,2) & \cdots & S(N,N) \end{pmatrix}$$

and

---

[4] $e_i = (0, 0, .., 1, \ldots, 0)'$ with 1 in the ith spot.

D Existence of a Solution

$$b = \begin{pmatrix} 1 \\ \rho \tilde{C}_1 \\ \rho \tilde{C}_2 \\ \cdot \\ \cdot \\ \cdot \\ \rho \tilde{C}_m \\ \rho S \end{pmatrix} \quad x = \begin{pmatrix} Q_0 \\ Q_1 \\ Q_2 \\ \cdot \\ \cdot \\ \cdot \\ Q_{N-1} \\ Q_N \end{pmatrix}.$$

Then (12.4), (12.5), (12.6) is the same as $Ax = b$ with these choices. Here A is an $(m+2) \times (N+1)$ matrix. Let us suppose that $w = (w_0, w_1, w_2, \ldots, w_m, w_{m+1})'$ satisfies (D.5). Then this means that (on dividing the right inequality in (D.5) by ρ)

$$w_0 \left[\frac{1}{\rho}\right] + w_1 \tilde{C}_1 + w_2 \tilde{C}_2 + \ldots + w_m \tilde{C}_m + w_{m+1} S > 0$$

and

$$w_0 + w_1 (S(N,0) - K_1)^+ + \ldots + w_m (S(N,0) - K_m)^+ + w_{m+1} S(N,0) \le 0$$
$$w_0 + w_1 (S(N,1) - K_1)^+ + \ldots + w_m (S(N,1) - K_m)^+ + w_{m+1} S(N,1) \le 0$$
$$\vdots$$
$$w_0 + w_1 (S(N,N) - K_1)^+ + \ldots + w_m (S(N,N) - K_m)^+ + w_{m+1} S(N,N) \le 0$$

We now perform the trades: If $w_0 > 0$, borrow $\frac{1}{\rho}$ from the bank, if $w_0 < 0$, invest $\frac{1}{\rho}$ in a bank; otherwise do nothing.

If for $1 \le i \le m$ we have $w_i > 0$, short sell the call with strike price K_i, while if $w_i < 0$, buy the call with strike price K_i; otherwise do nothing.

If we have $w_{m+1} > 0$, short sell the stock, while if $w_{m+1} < 0$, buy the stock; otherwise do nothing.

The result of constructing this portfolio is that at time $t=0$ (now) we have a profit

$$w_0 \left[\frac{1}{\rho}\right] + w_1 \tilde{C}_1 + w_2 \tilde{C}_2 + \ldots + w_m \tilde{C}_m + w_{m+1} S > 0,$$

and at $t = N$ we have

$$-w_0 - w_1 (S(N,\cdot) - K_1)^+ - \ldots - w_m (S(N,\cdot) - K_m)^+ - w_{m+1} S(N,\cdot) \ge 0.$$

D.2 An Application to the Problem 283

This says that we have no unfunded liabilities at $t = N$ in any state $j = 0, 1, 2, \ldots, N$. We thus have a (type 1) arbitrage opportunity. We thus conclude that if there is a solution to (D.5) then we have an arbitrage opportunity.

The outcome of all of this is the fact that (12.4)–(12.6) has a solution for $Q_j \geq 0$ for $j = 0, 1, 2, \ldots, M$ iff there are no arbitrage opportunities.

E

Some Generalizations

This note gives a generalization to some material in Section 12.5 of **van der Hoek's 1998 Method**. A key component was to find $Q_j \geq 0 : j = 0, 1, 2, \ldots, N$ satisfying the system:

$$1 = \sum_{j=0}^{N} Q_j \tag{E.1}$$

$$\exp(rT) \cdot S = \sum_{j=0}^{N} Q_j S(N, j) \tag{E.2}$$

$$\exp(rT) \cdot \tilde{C}_i = \sum_{j=0}^{N} Q_j \left[S(N, j) - K_i \right]^+ \quad \text{for } i = 1, 2, \ldots, m. \tag{E.3}$$

We select the $S(N, j)$ in such a way that a solution for the Q_j can be written down **explicitly**. In other words the $S(N, j)$ and Q_j are selected together.

E.1 Preliminary Observations

Lemma E.1. *Let $f \in C^2[0, \infty)$ and have compact support. Then*

$$f(x) = f(0) + x f'(0) + \int_0^\infty f''(y)(x - y)^+ \, dy \tag{E.4}$$

for all $x \geq 0$.

Proof. This follows from simple integration by parts. □

Lemma E.2. *Let f be piecewise linear on $[0, \infty)$. Then*

$$f(x) = f(0) + \sum_{i \geq 0} \alpha_i (x - x_i)^+, \qquad (\text{E}.5)$$

where $0 = x_0 < x_1 < x_2 < \ldots$ are the knots (kink points) of f, and $\alpha_i = m_i - m_{i-1}$, where $f'(x) = m_i$ on (x_i, x_{i+1}) for each $i \geq 1$ and $\alpha_0 = m_0$.

Proof. Both sides of (E.5) are piecewise linear, and it is easy to see that the two sides have the same value at $x = 0$ and the same slopes in each interval (x_i, x_{i+1}) with the choice suggested for the α_i.

Example E.3. Let $0 < K_1 < K_2 < K_3$. Let

$$f(x) = \frac{1}{K_2 - K_1}(x - K_1)^+ - \frac{K_3 - K_1}{(K_3 - K_2)(K_2 - K_1)}(x - K_2)^+ \\ + \frac{1}{K_3 - K_2}(x - K_3)^+. \qquad (\text{E}.6)$$

Then f is piecewise linear and

$$f(x) = \begin{cases} 0 & \text{if } x \leq K_1 \\ 1 & \text{if } x = K_2 \\ 0 & \text{if } x \geq K_3. \end{cases}$$

the payoff of a (call) butterfly spread.

Example E.4. With the special case $K_2 - K_1 = K_3 - K_2 = \Delta$ we obtain the representation

$$f(x) = \frac{1}{\Delta}\left((x - K_1)^+ - 2(x - K_2)^+ + (x - K_1)^+\right)$$

for

$$f(x) = \begin{cases} 0 & \text{if } x \leq K_1 \\ \frac{1}{\Delta}(x - K_1) & \text{if } x \in [K_1, K_2] \\ \frac{1}{\Delta}(K_3 - x) & \text{if } x \in [K_2, K_3] \\ 0 & \text{if } x \geq K_3. \end{cases}$$

E.2 Solution to System in van der Hoek's Method

We introduce some notation.

We assume $K_1 < K_2 < \ldots < K_m$. $\tilde{C}_j$ is the (market) price of a European call option expiring at time T with strike price K_j. $N = m - 1$, as we shall see.

$$\rho = \exp(r \cdot T) \quad \text{or} \quad P(0,T)^{-1}$$

We first write the choices for Q_j: For $j = 2, 3, \ldots, m-1$,

$$Q_j = \rho \left[\frac{1}{K_j - K_{j-1}} C_{j-1} - \frac{K_{j+1} - K_{j-1}}{(K_{j+1} - K_j)(K_j - K_{j-1})} C_j + \frac{1}{K_{j+1} - K_j} C_{j+1} \right] \tag{E.7}$$
$$> 0$$

by Example E.3, and

$$Q_j = 0 \quad \text{if} \quad j = 1, m \tag{E.8}$$

$$Q_{m+1} = \frac{\rho}{K_m - K_{m-1}} [C_{m-1} - C_m] \tag{E.9}$$

$$Q_0 = 1 - \sum_{j=1}^{m+1} Q_j. \tag{E.10}$$

The choices of S_j are

$$S_0 = \rho \left[\frac{S - \tilde{C}_1 - K_1 \frac{\tilde{C}_1 - \tilde{C}_2}{K_2 - K_1}}{1 - \rho \frac{\tilde{C}_1 - \tilde{C}_2}{K_2 - K_1}} \right] \tag{E.11}$$

$$S_j = K_j \quad \text{if} \quad j = 1, 2, \ldots, m \tag{E.12}$$

$$S_{m+1} = \rho \frac{\tilde{C}_m}{Q_{m+1}} + S_m. \tag{E.13}$$

Remark E.5. The choices made for Q_j in (E.7) are unique, but there are other possible choices for Q_0 and Q_{m+1} with (E.8). This needs some exploration. For example if K_j takes the form $b \cdot a^j$, then we might also like to have $Q_0 \approx Q_{m+1}$.

We can now establish the following facts, which follow if there are no arbitrage opportunities in the market containing cash, the asset, and the m European call options.

1. For $j = 0, 2, 3, \ldots, m-1, m+1$, $Q_j > 0$.

   For $j = 2, 3, \ldots, m-1$, this is true because Q_j/ρ is the price of a (call) butterfly spread, which must be positive. $Q_{m+1} > 0$ as $\tilde{C}_{m-1} > \tilde{C}_m$, and

   $$Q_0 = 1 - \frac{\rho}{K_2 - K_1}\left[\tilde{C}_1 - \tilde{C}_2\right] > 0,$$

   as it is ρ times the present value of and asset with pay-off

   $$f(S(T)) = \begin{cases} 1 & \text{if } 0 \leq S(T) \leq K_1 \\ \frac{K_2 - S(T)}{K_2 - K_1} & \text{if } K_1 \leq S(T) \leq K_2 \\ 0 & \text{if } S(T) \geq K_2. \end{cases}$$

   This is because Lemma E.1 implies that

   $$f(S(T)) = 1 - \frac{1}{K_2 - K_1}(S(T) - K_1)^+ + \frac{1}{K_2 - K_1}(S(T) - K_2)^+,$$

   and hence, assuming no arbitrage,

   $$\frac{1}{\rho} - \frac{1}{K_2 - K_1}\tilde{C}_1 + \frac{1}{K_2 - K_1}\tilde{C}_2 > 0.$$

2. It is now not difficult to show that (E.2) and (E.3) hold. For (E.3) we need to use $\tilde{C}_1 \geq S - \frac{K_1}{\rho}$, which implies that $K_1 \geq S_0$. Equation (E.3) is proved for $i = 1$ and then for $i > 1$.

E.3 Exercises

Exercise E.6. Use lemma (E.1) and (E.2) to explain how one can price a general European Contingent Claim in terms of the prices of European Call options. How could you use these ideas to obtain approximate pricing.

F

Yield Curves and Splines

In this appendix we discuss the approximation of yield curves by cubic splines. As in [48], let $m(t)$, $0 \leq t \leq T$ be the **discount function** of $[0, T]$. It describes the present ($t = 0$) value of $1 repayable in t years (if that is the unit of time). We shall seek an approximation of m on $[0, T]$ as a cubic spline. A cubic spline is a continuously differential piecewise cubic function, that is,

$$m(t) = a_i + b_i t + c_i t^2 + d_i t^3, \qquad t_i \leq t \leq t_{i+1}, \tag{F.1}$$

for $i = 0, 1, 2, \ldots, k$ where $0 = t_0 < t_1 < \ldots < t_k < t_{k+1} = T$.
The choice of t_i, $i = 1, 2, \ldots, k$ is up to the user. It is clear that one does not want too many **knots** t_i. In order to keep things simple one could take $t_i = \frac{i}{k+1} T$, that is equal-spaced, but this is not the best thing to do. If you expect the discount curve to have more curvature in $[0, T/2]$ than in $[T/2, T]$, then it may be advisable to have more knots in $[0, T/2]$. Let us assume that $t_1, t_2, \ldots, t_k$ are chosen and fixed throughout the discussion.

Some notation: If a function h is given as

$$h(t) = \begin{cases} f(t), & t < t_1 \\ g(t), & t > t_1 \end{cases}$$

and f, g are sufficiently smooth, then $h(t_1+)$, $h'(t_1+)$, $h''(t_1+)$ stand for respectively $g(t_1)$, $g'(t_1)$, $g''(t_1)$, and $h(t_1-)$, $h'(t_1-)$, $h''(t_1-)$ stand for $f(t_1)$, $f'(t_1)$, $f''(t_1)$. In order that m be a cubic spline we need

$$m(t_i+) = m(t_i-), \quad m'(t_i+) = m'(t_i-), \quad m''(t_i+) = m''(t_i-) \tag{F.2}$$

for $i = 1, 2, \ldots, k$. This is equivalent to

$$a_{i-1} + b_{i-1} t_i + c_{i-1} t_i^2 + d_{i-1} t_i^3 = a_i + b_i t_i + c_i t_i^2 + d_i t_i^3 \tag{F.3}$$

$$b_{i-1} + 2c_{i-1}t_i + 3d_{i-1}t_i^2 = b_i + 2c_i t_i + 3d_i t_i^2 \qquad (F.4)$$
$$2c_{i-1} + 6d_{i-1}t_i = 2c_i + 6d_i t_i \qquad (F.5)$$

for $i = 1, 2, \ldots, k$. As (F.3)–(F.5) are awkward to deal with, Litzenberger and Rolfo [48] proposed first to write m is a different way.

F.1 An Alternative representation of Function (F.1)

By (F.1) on $[t_1, t_2]$,

$$m(t) = a_1 + b_1 t + c_1 t^2 + d_1 t^3. \qquad (F.6)$$

We wish to write it as

$$m(t) = a_0 + b_0 t + c_0 t^2 + d_0 t^3 + A_1 + B_1(t-t_1) + C_1(t-t_1)^2 + F_1(t-t_1)^3. \qquad (F.7)$$

In fact if a_0, b_0, c_0, d_0 are known, then we can obtain a_1, b_1, c_1, d_1 from A_1, B_1, C_1, F_1 and vice versa. Let us equate powers of t in (F.6) and (F.7). Then

$$a_1 = a_0 + A_1 - B_1 t_1 + C_1 t_1^2 - F_1 t_1^3$$
$$b_1 = b_0 + B_1 - 2C_1 t_1 + 3F_1 t_1^2$$
$$c_1 = c_0 + C_1 - 3F_1 t_1$$
$$d_1 = d_0 + F_1,$$

which expresses a_1, b_1, c_1, d_1 in terms of A_1, B_1, C_1, F_1. Solving back, it is not hard to show that

$$F_1 = d_1 - d_0$$
$$C_1 = c_1 - c_0 + 3F_1 t_1$$
$$= c_1 - c_0 + 3(d_1 - d_0)t_1$$
$$B_1 = (b_1 - b_0) + 2C_1 t_1 - 3F_1 t_1^2$$
$$= (b_1 - b_0) + 2(c_1 - c_0)t_1 + 3(d_1 - d_0)t_1^2$$
$$A_1 = (a_1 - a_0) + B_1 t_1 - C_1 t_1^2 + F_1 t_1^3$$
$$= (a_1 - a_0) + (b_1 - b_0)t_1 + (c_1 - c_0)t_1^2 + (d_1 - d_0)t_1^3$$

after doing a little algebra, and this expresses A_1, B_1, C_1, F_1 in terms of a_1, b_1, c_1, d_1. So on $[t_0, t_1]$ we can write

$$m(t) = a_0 + b_0 t + c_0 t^2 + d_0 t^3 + [A_1 + B_1(t-t_1) + C_1(t-t_1)^2 + F_1(t-t_1)^3]D_1(t)$$

where $D_1(t) = 0$ for $t < t_1$ and $= 1$ for $t \geq t_1$.

We can repeat this argument, but instead of using $[t_0, t_1]$ and $[t_1, t_2]$ we use $[t_{i-1}, t_i]$ and $[t_i, t_{i+1}]$ for each i, and on $[t_i, t_{i+1}]$, $m(t)$ can be written

$$a_{i-1} + B_{i-1}t + c_{i-1}t^2 + d_{i-1}t^3 + A_i + B_i(t-t_i) + C_i(t-t_i)^2 + F_i(t-t_i)^3$$

as an alternative to $a_i + b_i t + c_i t^2 + d_i t^3$. Again we can express A_i, B_i, C_i, F_i in terms of a_i, b_i, c_i, d_i and vice versa. So on $[0, T]$

$$m(t) = a_0 + b_0 t + c_0 t^2 + d_0 t^3$$
$$+ \sum_{j=1}^{k} \left[A_j + B_j(t - t_j) + C_j(t - t_j)^2 + F_j(t - t_j)^3 \right] D_j(t)$$

where $D_j(t) = 0$ for $t < t_j$ and $D_j(t) = 1$ for $t \geq t_j$.

F.2 Imposing Smoothness

It is not difficult to show from (F.3)–(F.5) that

$$m(t_j+) - m(t_j-) = A_j$$
$$m'(t_j+) - m'(t_j-) = B_j$$
$$m''(t_j+) - m''(t_j-) = 2C_j$$

for $j = 1, 2, \ldots, k$. So if (F.2) holds then this is equivalent to $A_j = B_j = C_j = 0$, $j = 1, 2, \ldots, k$. We are left with

$$m(t) = a_0 + b_0 t + c_0 t^2 + d_0 t^3 + \sum_{j=1}^{k} F_j(t - t_j)^3 D_j(t), \qquad (\text{F.8})$$

for which (F.2) holds automatically.

F.3 Unknown Coefficients

We must now determine $(k + 4)$ unknowns

$$a_0, \ b_0, \ c_0, \ d_0, \ F_1, \ F_2, \ \ldots, \ F_k \qquad (\text{F.9})$$

Now since $m(0) = 1$, $a_0 = 1$, so we are left with $(k + 3)$ unknowns. These must be obtained from observations.

Remark F.1. We have used piecewise cubics. Piecewise functions of other types can be used. If we use polynomials we get more general splines. Fong and Vasicek [76] used piecewise exponentials.

F.4 Observations

We assume that the market provides us with bond prices. However, other instruments could also be used as in [48]. If a bond expires at T_n and pays coupon C at $T_1, T_2, \ldots, T_n$, then the present value of this bond is

$$P = \sum_{i=1}^{n} Cm(T_i) + Fm(T_n), \tag{F.10}$$

where F is the face value of the bond. We do not assume $T_1, T_2, \ldots, T_n$ are equally spaced, but in most cases they are. Of course there need be no connection between the T_j and the t_j.

We will have different choices of $\{C, F, T_1, T_2, \ldots, T_n, n\}$ for different bonds. We now substitute (F.8) into (F.10) to get

$$P = \sum_{i=1}^{n} C \left[1 + b_0 T_i + c_0 T_i^2 + d_0 T_i^3 + \sum_{j=1}^{k} F_j (T_i - t_j)^3 D_j(T_i) \right]$$

$$+ F \left[1 + b_0 T_n + c_0 T_n^2 + d_0 T_n^3 + \sum_{j=1}^{k} F_j (T_n - t_j)^3 D_j(T_n) \right].$$

We have assumed that $T_n \leq T$. So

$$P = [n \cdot C + F] + b_0 \left[C \sum_{i=1}^{n} T_i + F T_n \right]$$

$$+ c_0 \left[C \sum_{i=1}^{n} T_i^2 + F T_n^2 \right] + d_0 \left[C \sum_{i=1}^{n} T_i^3 + F T_n^3 \right] \tag{F.11}$$

$$+ \sum_{j=1}^{k} F_j \left[C \sum_{i=1}^{n} D_j(T_i)(T_i - t_j)^3 + F D_j(T_n)(T_n - t_j)^3 \right]$$

We can write (F.11) as

$$y = b_0 \xi + c_0 \eta + d_0 \zeta + \sum_{j=1}^{k} F_j z_j \tag{F.12}$$

where

$$y = P - nC - F$$

$$\xi = C \sum_{i=1}^{n} T_i + F T_n$$

$$\eta = C \sum_{i=1}^{n} T_i^2 + FT_n^2 \qquad (F.13)$$

$$\zeta = C \sum_{i=1}^{n} T_i^3 + FT_n^3$$

$$z_j = C \sum_{i=1}^{n} D_j(T_i)(T_i - t_j)^3 + FD_i(T_n)(T_n - t_j)^3$$

which are computable for each bond that we observe in terms of $C, F, T_1, T_2, \ldots, T_n$.

F.5 Determination of Unknown Coefficients

We determine the unknown coefficients $b_0, c_0, d_0, F_1, F_2, \ldots, F_k$ by linear regression. Suppose we have the price of N bonds from the market. For each we calculate the quantities in equation (F.13)

$$y^l, \xi^l, \eta^l, \zeta^l, z_1^l, z_2^l, \ldots, z_k^l$$

for $l = 1, 2, \cdots, N$. To simplify notation let us write

$$\mathbf{x}^l = (x_1^l, x_2^l, \ldots, x_M^l) \equiv (y^l, \xi^l, \eta^l, \zeta^l, z_1^l, z_2^l, \ldots, z_k^l)$$

where $M = k + 3$, and $\alpha = (\alpha_1, \alpha_2, \ldots, \alpha_M) \equiv (b_0, c_0, d_0, F_1, F_2, \ldots, F_k)$; then (F.12) is

$$y^l = \sum_{j=1}^{M} \alpha_j x_j^l, \quad l = 1, 2, \ldots, N$$

ideally! We choose α to minimize

$$J = \sum_{l=1}^{N} w_l \left[y^l - \sum_{r=1}^{M} \alpha_r x_r^l \right]^2 \qquad (F.14)$$

where $w_l > 0$ for $l = 1, 2, \ldots, N$ and $\sum_{l=1}^{N} w_l = 1$. For a least squares approximation we would normally set $w_l = 1/N$ for all l, but we have the flexibility to give more weight to certain data. To find $\hat{\alpha}$ optimal, we solve:

$$\frac{\partial J}{\partial \alpha_j} = 0, \quad j = 1, 2, \ldots, M \qquad (F.15)$$

which is the same as

$$-2\sum_{i=1}^{N} w_l \left[y^l - \sum_{r=1}^{M} \hat{\alpha}_r x_r^l \right] x_j^l = 0, \quad \text{for } j = 1, 2, \ldots, M \tag{F.16}$$

or

$$\sum_{r=1}^{M} \hat{\alpha}_r \sum_{l=1}^{N} w_l x_r^l x_j^l = \sum_{l=1}^{N} w_l y^l x_j^l \tag{F.17}$$

for $j = 1, 2, \ldots, M$. This is a matrix equation for $\hat{\alpha}$ which is

$$V\hat{\alpha} = \beta \tag{F.18}$$

with $\beta = (\beta_1, \beta_2, \ldots, \beta_M)$ and $\beta_j = \sum_{l=1}^{N} w_l y^l x_j^l$ and $V = (V_{ij})$, and

$$V_{ij} = \sum_{l=1}^{N} w_l x_i^l x_j^l \tag{F.19}$$

We need to show that V is **invertible**. In fact if V is positive definite, then V is invertible. This will be the case if

$$\xi^T V \xi \geq 0 \quad \text{for any } \xi \in \mathbf{R}^M$$

and $\xi^T V \xi = 0$ implies $\xi = 0$. But

$$\xi^T V \xi = \sum_{i,j=1}^{M} V_{ij} \xi_i \xi_j = \sum_{l=1}^{N} w_l \left(\sum_{i=1}^{M} \xi_i x_i^l \right)^2 \geq 0$$

and $\xi^T V \xi = 0$ implies $\sum_{i=1}^{M} \xi_i x_i^l = 0$ for $l = 1, 2, \ldots, N$ which implies $\sum_{i=1}^{M} \xi_i \mathbf{x}_i = 0$ where $\mathbf{x}_i = (x_i^1, x_i^2, \ldots, x_i^N) \in \mathbf{R}^N$. This will imply that $\xi_i = 0$ for all i if the $\mathbf{x}_i$, $i = 1, 2, \ldots, M$ are linearly independent. This implies that we need $M \leq N$. So we *need* at least M bond prices, that is at least $k + 3$ bond prices to determine the unknowns. However it could happen that with $N \geq k + 3$ that the $\mathbf{x}_i$ are still linearly dependent (in which case V is not invertible), but if the N bond prices contain N "independent" pieces of information this is not likely. In any case once we choose $N \geq k + 3$ bond prices we need to check that V has an inverse, else we have to get some extra bond prices.

We next outline some computation details.

$$\Phi = \begin{bmatrix} \xi^1 & \xi^2 & \cdots & \xi^N \\ \eta^1 & \eta^2 & \cdots & \eta^N \\ \zeta^1 & \zeta^2 & \cdots & \zeta^N \\ z_1^1 & z_1^2 & \cdots z_1^N \\ \vdots & \vdots & & \vdots \\ z_k^1 & z_k^2 & \cdots & z_k^N \end{bmatrix}$$

$$\Lambda = \mathbf{diagonal}\,[w_1, w_2, \ldots, w_N] \tag{F.20}$$

$$V = \Phi\,\Lambda\,\Phi^T$$

$$\beta = [\beta_1, \beta_2, \ldots, \beta_M]^T$$

$$V^{-1}\beta = \begin{bmatrix} b_0 \\ c_0 \\ d_0 \\ F_1 \\ \vdots \\ F_k \end{bmatrix} \tag{F.21}$$

We can then (if we wish) calculate the a_i, b_i, c_i, d_i in equation (F.1) as follows:

$$a_i = a_{i-1} - F_i t_i^3$$
$$b_i = b_{i-1} + 3F_i t_i^2$$
$$c_i = c_{i-1} - 3F_i t_i$$
$$d_i = d_{i-1} + F_i$$
$$a_0 = 1 \quad \text{and} \quad b_0, c_0, d_0 \quad \text{from (F.21)}$$

J.H. McCulloch [50] mentions that we do not have bond prices for each of the N bonds, but rather bid- and ask-spreads, (see [50, pages 20–21]).

F.6 Forward Interest Rates

If $f(t)$ is the instantaneous forward interest rate, then

$$m(t) = \exp\left[-\int_0^t f(s)ds\right], \tag{F.22}$$

which implies that

$$f(t) = -\frac{m'(t)}{m(t)}, \tag{F.23}$$

from which

$$\hat{f}(t) = -\frac{[2b_i c_i t + 3d_i t^2]}{[a_i + b_i t + c_i t^2 + d_i t^3]} \tag{F.24}$$

when $t_i \leq t \leq t_{i+1}$.

F.7 Yield Curve

If $y(t)$ is the yield for the period $[0, t]$ then

$$m(t) = \exp\left[-ty(t)\right] \tag{F.25}$$

so

$$y(t) = -\frac{1}{t}\log_e[m(t)] \tag{F.26}$$

so

$$\hat{y}(t) = -\frac{1}{t}\log_e\left[a_i + b_i t + c_i t^2 + d_i t^3\right] \tag{F.27}$$

for $t_i \leq t \leq t_{i+1}$.

F.8 Other Issues

1. Given the error in spline approximation to the discount curve, we can compute the error for the forward interest rates and the yield curve. This analysis is discussed in McCulloch [50].

2. In choosing the size of k (the number of "knots") we need to take into account the following. If k is too small, we cannot get a good fit. If k is too close to N there are problems alluded to above. McCulloch suggests $k \approx N^{\frac{1}{2}}$.

3. The appropriate estimation depends on the observables. We have assumed one price for N different bonds. McCulloch has spreads on N different bonds, and so the analysis differs a little from that given above.

4. Other features can be incorporated. McCulloch [51] discusses tax-adjusted yield curves and Litzenberger and Rolfo [48] also discuss tax effects on government bonds.

References

1. M. Abramowitz and I.A. Stegun, *Handbook of Mathematical Functions with Formulas, Graphs, and Mathematical Tables*, Dover, New York, 1965.
2. S. Anthony, *Foreign Exchange in Practice: The New Environment*, Palgrave Macmillan, New York, 2002.
3. S. Barle and N. Cakici, *Growing a Smiling Tree*, The Journal of Financial Engineering **7** (1998), no. 2, 127–146.
4. S. Beckers, *Standard Deviations Implied in Option Prices as Predictors of Future Stock Price Variability*, Journal of Banking and Finance **5** (1981), 363–381.
5. P. Bjerksund and G. Stensland, *Implementing the Black-Derman-Toy Interest Rate Model*, Journal of Fixed Income **6** (1996), no. 2, 67–75.
6. T. Bjork and B.J.Christensen, *Interest Rate Dynamics and Consistent Forward Rate Curves*, Mathematical Finance **9** (1999), 323–348.
7. F. Black, E. Derman, and W. Toy, *A One Factor Model of Interest Rates and its Application to Treasury Bond Options*, Financial Analysts Journal **46** (1990), 33–39.
8. F. Black and M. Scholes, *The Pricing of Options and Corporate Liabilities*, Journal of Political Economy **81** (1973), 637–654.
9. P. Boyle and K. S. Tan, *Lure of the Linear*, Risk **7** (1994), 43–46.
10. G. Brown and K. J. Toft, *Constructing Binomial Trees from Multiple Implied Probability Distributions*, The Journal of Derivatives **7** (1999), no. 2, 83–100.
11. Ren-Raw Chen, *Understanding and Managing Interest Risks*, World Scientific Pub Co Inc, Singapore, 1996.
12. M. Chesney and H. Loubergé, *Les Options de Change*, Presses Universitaires de France, Paris, 1992.
13. N. Chriss, *Transatlantic Trees*, RISK Magazine (1996).
14. N. A. Chriss, *Black Scholes and Beyond: Option Pricing Models*, McGraw-Hill, New York, 1996.
15. Kai Lai Chung, *A Course in Probability Theory*, 3rd ed., Academic Press, San Diego, USA, 2001.
16. Vašek Chvátal, *Linear Programming*, W.H.Freeman and Co., New York, 1983.
17. T. E. Copeland and V. Antikarov, *Real Options: A Practitioner's Guide*, 1st ed., Thomson Texere, New York, Feb. 2001.
18. J. C. Cox, S. A. Ross, and M. Rubinstein, *Option Pricing: A Simplified Approach*, Journal of Financial Economics **7** (1979), no. 3, 229–263.

19. J. C. Cox and M. Rubinstein, *Options Markets*, Prentice Hall, Englewood Cliffs, NJ, 1985.
20. R-A. Dana and M. Jeanblanc, *Financial Markets in Continuous Time*, Springer Finance, New York, 2003.
21. R. E. Dattatreya and K. Hotta, *Advanced Interest Rate and Currency Swaps*, Irwin Professional Publishing, Burr Ridge, IL, 1993.
22. M. H. A. Davis, *Optimal Hedging with Basis Risk*, Working Paper (2000).
23. O. de la Grandville, *Bond Pricing and Portfolio Analysis*, MIT Press, Boston, MA, 2001.
24. E. Derman and I. Kani, *Volatility Smile and Its Implied Tree*, (1994).
25. A. K. Dixit and R. S. Pindyck, *Investment under Uncertainty*, Princeton University Press, Princeton, NJ, 1994, Reviewed by E. Schwartz in the Journal of Finance 49 (1994), 1924-1926.
26. D. Duffie, *Futures Markets*, Prentice Hall, Upper Saddle River, NJ, 1989.
27. R. Elliott and J. van der Hoek, *Pricing Non-Tradeable Assets: Duality Methods*, in R. Carmona (ed) "Indifference Pricing" Princeton University Press 2005.
28. _____, *Pricing Claims on Non Tradeable Assets*, Contemporary Mathematics **351** (2004), 103–114.
29. R. J. Elliott and P. E. Kopp, *Mathematics of Financial Markets*, 2nd ed., Springer Verlag, New York, 2004.
30. H. U. Gerber and W. Neuhaus, *Life Insurance Mathematics*, 3rd ed., Springer Verlag, New York, 1997.
31. R. Geske, *The Valuation of Compound Options*, Journal of financial economics **7** (1979), 63–81.
32. V. Henderson, *Valuation of Claims on Non-Traded Assets using Utility Maximization*, EFMA 2001 Lugano Meetings, May 2001.
33. T. S. Y. Ho and S. Lee, *Term Structure Movements and Pricing Interest Rate Contingent Claims*, Journal of Finance **41** (1986), no. 5, 1011–1029.
34. _____, *Duration, Futures Pricing and Immunization with Interest Rate Futures*, Working Paper **41** (1988).
35. _____, *Interest Rate Futures Options and Interest Rate Options*, The Financial Review **25** (1990), 345–370.
36. J. Hull and A. White, *Efficient Procedures for valuing European and American Path-Dependent Options*, Journal of Derivatives **Fall** (1993), 21–31.
37. J. C. Hull, *Options, futures and Other Derivatives*, 5th ed., Prentice Hall, London, 2003.
38. W. Hurlimann, *On Binomial Models of the Term Structure of Interest Rates*, Proceedings of the 5th AFIR International Colloquium, Brussels 1995, 833–858.
39. J. C. Jackwerth, *Generalized Binomial Trees*, The Journal of Derivatives **5** (1997), no. 2, 7–17.
40. L. L. Jacque, *Management and Control of Foreign Exchange Risk*, Kluwer, 1996.
41. F. Jamshidian, *Forward Induction and Construction of Yield Curve Diffusion Models*, The Journal of Fixed Income (1991), 62–74.
42. R. Jarrow (ed.), *Over the Rainbow : Development in Exotic Options and Complex Swaps*, Risk Publications, London, UK, 1995.
43. R. Jarrow and M. Turnbull, *Derivative Securities*, 2nd ed., South-Western Publishing, Philadelphia, PA, 2000.
44. B. A. Jensen, *Binomial Models for the Term Structure of Interest*, (2000).
45. H. A. Latane and R. J. Rendleman, *Standard Deviations of Stock Price Ratios Implied in Option Prices*, The Journal of Finance **31** (1976), no. 2, 369–381.

46. K. G. Lim and D. Zhi, *Pricing Options Using Implied Trees: Evidence from FTSE-100 Options*, The Journal of Futures Markets **22** (2002), no. 7, 601–626.
47. R. H. Litzenberger and J. Rolfo, *An International Study of Tax Effects on Government Bonds*, The Journal of Finance **39** (1984), no. 1, 1–22.
48. _____, *An International Study of Tax Effect on Government Bonds*, The Journal of Finance **39** (1993), no. 1, 1–22.
49. S. Mayhew, *Implied Volatility - Literature Survey*, Financial Analyst Journal (1995), 8–20.
50. J. H. McCulloch, *Measuring the Term Structure of Interest Rates*, Journal of Business **44** (1971), no. 1, 19–31.
51. _____, *On Tax Adjusted Yield Curves*, The Journal of Finance **30** (1975), 811–830.
52. _____, *The Tax-Adjusted Yield Curve*, Journal of Finance **30** (1975), no. 3, 811–830.
53. R. Merton, *The Theory of Rational Option Pricing*, Bell Journal of Economics and Management Sciences **4** (1973), 141–183.
54. R. C. Merton, *On the Pricing of Corporate Debt: The Risk Structure of Interest Rates*, Journal of Finance **2** (1974), 449–470.
55. I. G. Morgan and E. H. Neave, *A Discrete Time Model for Pricing Treasury Bills, Forward, and Futures Contracts*, Astin Bulletin **23** (1993), no. 1, 3–22.
56. M. Musiela and T. Zariphopoulou, *A Valuation Algorithm for Incomplete Markets*, Conference of Bachelier Society of Finance, Crete, 2002.
57. S. Natenberg, *Option Volatility and Pricing*, 2nd ed., Irwin, 1994.
58. C. R. Nelson and A. F. Siegel, *Parsimonious Modeling of Yield Curves*, Journal of Business **60** (1987), no. 4, 473–489.
59. H. H. Panjer (ed.), *Financial Economics: with Applications to Investments, Insurance, and Pensions*, Society of Actuaries Foundation, 1999.
60. H. Pedersen, E. Shiu, and A. Thorlacius, *Arbitrage-Free Pricing of Interest Rate Contingent Claims*, Transactions of the Society of Actuaries **41** (1990), 231–279.
61. A. Pelsser and T. Vorst, *The Binomial Model and the Greeks*, Journal of Derivatives **1** (1994), no. 3, 45–49.
62. S. Pliska, *Introduction to Mathematical Finance: Discrete Time Models*, Blackwell, Oxford, U.K., Malden, MA., 1997.
63. Jr. R. J. Rendleman and B. J. Bartter, *Two-State Option Pricing*, The Journal of Finance **34** (1979), no. 5, 1093–1110.
64. P. Ritchken and K. Boenawan, *On Arbitrage-Free Pricing of Interest Rate Contingent Claims*, Journal of Finance **45** (1990), no. 1, 259–264.
65. M. Rubinstein, *Implied Binomial Trees*, The Journal of Finance **49** (1994), no. 3, 771–818, Papers and Proceedings Fifty-Fourth Annual Meeting of the American Finance Association, Boston, Massachusetts, January 3-5.
66. _____, *As Simple as One, Two, Three*, RISK (1995).
67. K. Sandmann and D. Sondermann, *A Term Structure Model and the Pricing of Interest Rate Derivatives*, Universitaet Bonn: Discussion Paper B-180 (1991).
68. _____, *A Term Structure Model and the Pricing of Interest Rate Derivatives*, The Review of Futures Markets **12** (1993), no. 2, 391–423.
69. E. Schwartz, *Book Review: Investment Under Uncertainty*, Journal of Finance **49** (1994), no. 5, 1924–1926.
70. W. E. Sharpe, *Investments*, Prentice-Hall Inc., Englewood Cliffs, N.J., 1978.
71. M. Sherris, *Money and Capital Markets: Pricing Yields and Analysis*, 2nd ed., Allen and Unwin Pty Ltd, 1997.

72. G. Sick, *Book Review: Real Options (Book) by L. Trigeorgis*, Journal of Finance **51** (1996), no. 5, 1974–1977.
73. J. E. Smith and R. F. Nau, *Valuing Risky Projects: Option Pricing Theory and Decision Analysis*, Management Science **41** (1995), no. 5, 795–816.
74. A. Street, *Stuck up a Ladder*, Risk (1992).
75. L. Trigeorgis, *Real Options: Managerial Flexibility and Strategy in Resource Allocation*, MIT Press, 1996, Reviewed by G. Sick in the 51 (1996), 1974-1977.
76. O. A. Vasicek and H. G. Fong, *Term Structure Modeling using Exponential Splines*, Proceedings of American Finance Association Meeting, December 1981.
77. J. von Neumann and O. Morgenstern, *The Theory of Games and Economic Behaviour*, 1947.
78. W. L. Winston, *Operations research - applications and algorithms*, 3 ed., Duxbury Press, Belmont, California, 1994.
79. P. G. Zhang, *Exotic Options: A Guide to Second Generation Options*, World Scientific Publishing, Singapore, 1997.

Index

AA market, 176
abandonment, 235
arbitrage, 1
    type one, 24
    type two, 24
Arrow-Debreu security, 46
attainable claims, 250

backward induction pricing formula, 67
bank bill swap rate (BBSW), 176
Barle and Cakici approach, 145
barrier option, 99
    barrier monitoring, 100
    customization, 100
    down-and-in put, 103
    down-and-out call, 103
    knock-in, 99
    knock-out, 99
    specifications, 100
    up-and-in, 101
    up-and-in call, 104
    up-and-out, 100
    up-and-out call, 105
BBSW, 176
Bernoulli trials, 237
Berry - Esséen Theorem, 245
Berry-Esséen Theorem, 77
beta, 23
bid-ask spreads, 250
binomial asset pricing model, 13
    multiperiod, 65, 76
    one step, 21
    one-step, 75
binomial distribution function, 75

complementary, 74, 237
bisection method, 194, 274
Black and Scholes formula, 26
Black and Scholes formula, 76
Black, Derman and Toy model, 58, 193
Bob Arnold, 183
bond
    zero-coupon/T-zero, 56
bonds
    defaultable, 205
Bretton-Woods agreement, 45
butterfly spread, 286

CAD, 45
call-put parity, 27
capital asset pricing model, 24
certainty equivalence, 214
clearing house, 90
complete the market, 251
compound options, 79
Copeland and Antikarov, 223
Cox-Ross-Rubinstein
    model, 25
    model with dividends, 129
    multiperiod model, 68
    option pricing formula, 73
creditworthy, 171, 181
cubic splines, 289

delta, 121
derived asset, 14
Derman-Kani method, 137
divisible market, 16
duality theorem, 256, 261

Index

early-exercise premium, 32, 98
electricity supply problem, 242
ex-dividend date, 127
exchange rates, 45
   direct/American quotation, 45
   futures contract, 52
   inverse/European quotation, 45
exercise date, 5
expected utility, 214
expiration date, 5

Farkas' lemma, 279
first fundamental theorem of finance, 257
forward
   contract, 3, 41, 89
   exchange rate contract, 50
   price, 41
   rate, 171
futures contract, 4, 90
   default, 94
   equivalence with forward price, 95

gamma, 123
generalized binomial trees, 168
Greeks, 126

Heath-Jarrow-Morton framework, 172
hedge ratio, 82
hedging, 81
Ho and Lee model, 57, 172, 184
Hull and White method, 115
Hull-White interest rate model, 184

implied binomial tree, 153
implied volatility surface, 145
implied volatility trees, 135
in-the-money, 10
indifference price, 216
   asking, 216
   bid, 216
interest rate, 14
   derivative, 55
   parity (covered), 51
interpolation, 150

Jackwerth's Extension, 168
Jamshidian's forward induction
   formula, 67, 69

Kuhn-Tucker theorem, 253

law of one price, 2
linear regression, 36

MAD, 223
margin account, 52, 90
margin call, 90
market players
   arbitrageurs, 9
   hedgers, 9
   speculators, 9
marking to market, 91
martingale, 160
model-independent formulae, 27
Morgan and Neave model, 191

Nelson-Siegel approach, 183
Newton-Raphson method, 194, 273
nonrecombining binomial tree, 109
normal distribution function, 244

one, two, three algorithm, 155
option, 5
   American, 5, 97
   American call, 31
   American put, 32
   Asian/average rate, 109
   Bermuda, 7
   binary, 79
   booster, 107
   call, 5
   chooser/as you like it, 78
   compound, 79
   contingent premium/pay later, 43
   European, 5
   exchange traded, 7
   exotic, 43
   floating strike, 112
   forward start, 79
   ladder, 119
   lookback, 113
   lower bounds, 30
   partial barrier, 107
   perpetual, 5
   put, 5
   real, 210, 235
   strike, 7
   style, 7

vanilla/plain, 43
over the counter/OTC, 49

payer-swap, 175
Pedersen, Shiu and Thorlacius model, 189
present value, 56
put-call parity, 27

quadratic linear programming, 161

real options, 209
  growth/to expand, 212
  to abandon, 213, 223
  to contract, 212, 224
  to default, 211
  to defer, 211
  to expand, 225, 226
  to shut down, 212
receiver-swap, 175
relative pricing, 14, 19
replicating portfolio, 19
resettlement, 91
return, 21
rho, 125
risk avertors, 9
risk-adjusted, 23
risk-neutral, 23
  expectation, 19
  probability, 19, 21
Rubinstein's 1994 method, 161

secant method, 195
second fundamental theorem
  of finance, 264
self-financing, 81
short selling, 16
splines, 289
spot rate, 172
state price, 47, 69
subreplication, 249
superreplication, 249
swap rate, 175

T-forward exchange rate, 50
theta, 124
tradeable asset, 13
transaction costs, 266

underlying, 6
USD, 45

van der Hoek's 1998 method, 162, 285
vega, 125, 273
volatility
  absolute and proportional, 195
  estimation, 269
  historical, 270
  implied, 272
  smile, 150

yield to maturity, 171